D1229926

Concession to the Improbable

GEORGE GAYLORD SIMPSON

George Gaylord Simpson

Concession to the Improbable

An Unconventional Autobiography

New Haven and London

Yale University Press

1978

Published in Great Britain, Europe, Africa, and Asia (except Japan) by Yale University Press,
Ltd., London. Distributed in Latin America by Kaiman & Polon, Inc., New York City; in
Australia and New Zealand by Book & Film Services, Artarmon, N.S.W., Australia; and in Japan
by Harper & Row, Publishers, Tokyo Office.

Library of Congress Cataloging in Publication Data

Simpson, George Gaylord, 1902—
    Concession to the improbable.

    Includes index.
    1. Simpson, George Gaylord, 1902—   2. Paleontologists—United States—Biogra-
phy. I. Title. QE707.S55A3   560'.92'4 [B]   77-20246 ISBN 0-300-02143-7

Dedicated with love to Anne, Helen, Gay, Joan, and Elizabeth

"All life is a concession to the improbable"

— comment ascribed to Thomas Heard in *South Wind*, by Norman Douglas

Contents

Thanks

As often before, but even more than usual, my wife, Anne Roe, urged and cheered me on as I wrote this book. She also read all the manuscript and detected rough spots and mistakes. Carol Eley performed the seemingly impossible feat of simultaneously earning a law degree and transforming my nearly indecipherable handwriting into neat typescript. Yale University Press enthusiastically supported this project when I was close to aborting it, and Ellen Graham and Catherine Iino skillfully edited the manuscript. The text itself indicates that I am indebted for my subject matter to some hundreds, if not thousands, of people, just a few of whom are named.

Foreword: What It's About

As the noon sun moved higher the days warmed and the last crusts of snow disappeared from the north sides of the clumps of sage brush. Finally came the balmy spring day when feet could be emancipated from confining shoes and legs from ugly long black stockings. Then I could run free and barefooted out onto the prairie that began where Denver's Milwaukee Street ended, a few yards beyond our house. I ran in narrow ruts where dust oozed delightfully between my toes, and I somersaulted in shallow pans, mud-cracked circles scattered here and there.

Even then I knew what those ruts and pans were: the trails and wallows made by the once great herds of buffalo. (None of us then had ever heard the word *bison*.) Gone. Almost all gone, but not quite. A tiny dejected remnant survived behind wire fences in a far corner of City Park. I gazed at them with awe, and I followed the traces of their mighty forebears with delight.

On those first spring runs I would plop down to see the tiny yellow violets we called johnny-jump-ups or perhaps to pluck one of the white sand lilies, blossoming without stems directly on the soil. I had begun to discover how full of interesting things is the world. Wonder increased as I learned in school that there are incredibly broader horizons, and that books are one way to explore them.

My pleasure in learning was too great to be concealed. Having been compared unfavorably to me by the principal of what was then Clayton School (later Stevens, in honor of that same principal), two tough boys waylaid me after school and held me while their younger brother methodically beat me up. I accepted the judgment that I was eccentric and the circumstance that eccentricity is punished, but I did not change my ways.

Many years and sights and emotions later I was a man of twenty-one and in France for the first time. There were no ghostly buffalo herds. There were not, in Île-de-France, any of the great mountains that I had come to love even more than the prairies. But there was an almost blinding and deafening flood of new sights and sounds. Smells, too, and I miss here a word like *blind* and *deafen* for "make anosmic," a linguistic poverty

1

signaling the lesser interest of that sense to humans. I thought there should be a potpourri of a book called simply *Interesting Things*, and I spent some reveries selecting topics I would then have included in such a work: "Religious iconography of the Middle Ages" — although the Christian myths were not altogether fascinating, their sculptured depictions in Notre Dame and Chartres were. Architecture, too, and, yes, with a touch of Gothic, but mostly Romanesque. (I had gazed at Vézelay by moonlight with a French sculptor-architect.) The Egyptian language. (I had already learned to read simple hieroglyphs and could practice in the Louvre.) Italian grand operas (I had heard most of them as an usher in the Chicago Opera, and the memory came back in these reveries). And much, much more.

My tastes have changed considerably, as tastes fortunately do. "Fortunately" because change brings variety. The change is not total. Sculpture still, and still the serene kings and queens of Chartres, but now also, and even more, the grotesqueries of Sepik or Makonde wood carving. Architecture: the mosques of Cairo excel. Language, just now Hawaiian or Fijian in preference to Egyptian. Painting: A positive dislike for Ingres and David accompanies a liking for Picasso. Italian opera has not become a real dislike, but is merely too silly to attend to. Only recently I discovered Schubert's great ninth symphony in C major, and wonder why you hear only his melodious but lesser *Unfinished*.

This book is written because the world is indeed so full of a number of things, and in a sense this is the *Interesting Things* of my distant fantasies. Nevertheless I shall have little occasion to refer again to the subjects of the projected chapters so far mentioned. This is to be one book, and it is harder to write one book than ten. Choices must be made. Many of the interesting things must be excluded. A distinction must be made between private and public interest — not that either must be eliminated, but evaluation must occur with the distinction in mind. Most of the things in my life that might possibly have some public interest have happened since those first days in Île-de-France.

Now I believe that I have what this book is to be about fairly boxed in. It's about me, but, from that vantage point, about much else as well. It's about things that I have done and seen and thought and that I consider interesting and in some parts important. However, I judge them to be interesting and in those parts important by criteria other than the mere fact that they concern me.

The things that I judge to be most important are so because they involve quite basic problems of science and philosophy. What is the history of life? What are the causes and patterns of evolution? How has the history of continents affected the distribution of animals? What is the nature of the human species? What of ethics and gods? Everyone must agree that those are important questions, whatever may be the ultimate judgment on my contributions to possible answers. The things that I judge to be less generally important but still interesting, excluding for present purposes excursions into iconography, linguistics, art, and such nonbiographical matters, involve personal experiences and interests and especially expeditions and travels.

Now that I have laid down the ground rules, I only add that I reserve the right to break or at least to bend them a bit from time to time. Strictly applied, they might exclude such matters as where my Sunday school class was held or what instruments have suffered in my musical careers, and I do intend to include a few such inconsequential items.

1

One Day

Sufficient unto the day . . . — Saint Matthew

We had been camped at Uruyén in the Kamarata Valley for a long time. Our Venezuelan geologists had made a good study of the region. The astronomers had so far failed to locate it precisely. Anne had made the first collection of mammals from that locality. I had made a demographic and ethnographic study of the Kamarakoto tribe, had made the first record of their dialect, and had obtained the long and fascinating legend of Maichak. Most exciting of all, Jimmie Angel had flown us over and past the falls he had discovered, had demonstrated against universal skepticism that they do exist, and had convinced us against all probability that they are by a good margin the highest waterfalls in the world. For days we had wanted to go on to completion of our effort by a visit to Santa Elena, but had been thwarted by absurdity, chicanery, noncooperation, stupidity, and indifference in some quarters that had so far almost evenly balanced the sense, honesty, cooperation, intelligence, and concern of the geologists, the Angels, and a few others.

That morning Jimmie Angel and his rickety one-motor plane still had some taxi work to do. An engineer and a geologist had to be hopped down the Akanan River. Another geologist had to be lifted to the 1,100-meter camp. Finally at 11:15 we could take off from the hastily cleared dirt strip near Uruyén. Jimmie was pilot; also on board were his wife, Marie; Victor López (one of the Venezuelan geologists); my wife, Anne; I; and a Kamarakoto Indian whom we called Antonio. (Kamarakotos never tell their real names, because that is dangerous.) Antonio had never flown before and his far from impassive face was a show of pride and terror. The plane was the only vehicle, wheeled or not, ever

4

seen by his tribe. We left him, at his request, at the little mission station of Luepa at the northern end of the Gran Sabana, a vast partly grassy area amidst the tropical forests of southeastern Venezuela.

The sabana is hilly, with sudden drops and rises from one level to another. Numerous rivers meander tortuously across grassy plots, then drop suddenly to other levels or flow into stretches of forest, for most of the hills are forested. Patches, perhaps a mile across, of the densest sort of tropical jungle end abruptly without the slightest transition and give way to smooth and rolling treeless, grassy plains. There are high mountains on every side, but of these we got only glimpses through the heavy storm clouds. As we skirted some smaller storms, at times it was raining on one wing of the plane and not on the other. The country became rougher and rougher until to our lay eyes, at least, there were no possible landing fields. Then a great bank of unskirtable clouds cut off our whole course.

We swung along the front of clouds until Jimmie found a small window and dived through to fly below it. Still we could only see the ground occasionally, and what we saw was not very comforting. As it cleared a bit, an occasional Indian hut appeared, but nothing that looked like Santa Elena. Somehow we had all — even Marie — assumed that Jimmie had been to Santa Elena. In fact he had never said so, and he neither had been there nor really knew the way. There was no map; mapping the region was one of the assignments for which the Venezuelan party back at Uruyén had been sent and would later proceed to Santa Elena.

Time passed. The gasoline gauge moved toward zero. There could no longer be any doubt: we had missed Santa Elena. There was no conversation, but we all had our thoughts. We were frightened, and would have been fools not to have been. Perhaps the circumstances should be made a little clearer. In spite of Jimmie's care, the plane was old, rickety, and barely airworthy. It had a minimum of instruments and no radio. We were lost in stormy weather in a region never mapped and rarely visited. It was the same region into which a pilot named Redfern had flown not long before, never to be seen again. In the clouds all around us were the strange mountains called *tepuis* (their Indian name): flat-topped eminences rising in vertical cliffs several thousand feet from the lowland sabana and valleys. Roraima, Conan Doyle's "Lost World," is a tepui that towers above Santa Elena. Flying into one of those — a possibility every time we entered a cloud — might have been more merciful than the perhaps more lingering fate of a forced landing in the forest.

Anne's thoughts, as told later, were probably the most practical. She had planned to collect mammals around Santa Elena and had some snap traps with her. She was wondering how long we might keep alive by trapping rats. She was also regretting that the strongest painkiller in her medicine kit was codeine, and not enough of that.

When we finally did reach Santa Elena they told us that we had flown right overhead, hidden in the clouds, and had gone on, far south into Brazil.

Even at the time it was obvious that we had overshot, so Jimmie circled back, finally to a large river. I guessed it to be the Kukenan, a wild guess, but it turned out to be correct. Even if we had been sure, our troubles would not have been over, for Santa Elena is not on the Kukenan proper but up a tributary. And we didn't know whether the tributary was above or below where we hit the river that might have been the Kukenan. A difference of opinion arose, and we tried a little of each to no avail as the gasoline dwindled still further.

We saw an Indian village, just a couple of huts in a sabana, and Jimmie went down and circled it at low altitude several times. To our surprise, among those who swarmed out of the huts was a priest. On our next circle Jimmie landed, a ticklish moment, for of course no plane ever was there before and the bunch grass of the sabanas can flip a small plane right over. One of the great old-time, seat-of-the-pants bush pilots made another perfect landing. (I'm sure no one was more surprised than he when poor Jimmie died in a bed some years later.)

In a moment we were surrounded. The priest climbed into the plane, and I climbed out with the Indians who were crowded under the wings because it was pouring rain and even an Indian doesn't like to get his G-string (or, in Kamarakoto, *wayiku*) wet. (As a matter of fact these were Taurepanes and their word was *kamizha*.) So we learned that we were at Apiyai-Kupue on the north bank of the Kukenan and that the priest was from Santa Elena, two days distant as he traveled on foot.

Soon we took off again, after a terrific thump on a bunch of tuft grass but no harm done. The priest, a welcome guide and hitchhiker, was even more picturesque than most, long-robed, bare-headed, with a beard almost to his waist. As was natural, he knew his way only by footpath, and this we followed to Santa Elena. Only a superb bush pilot could fly a plane along a winding trail!

Over a final ridge, we came into the valley of Santa Elena and a sabana possible for landing. Jimmie made a low pass, but at the last minute gunned up again abruptly, circled, and finally set us down. He came back from the pilot's seat crimson with anger and using language fortunately incomprehensible to the good father, who spoke only Taurepán and Spanish. It seems that at the last minute the padre had leaned over to point the way, and his whiskers had come adrift and blown across Jimmie's face, completely obscuring his vision. Jimmie explained that he couldn't even have made an instrument landing, overlooking the fact that there were no landing instruments anyway. Jimmie hurriedly took off again, still into the storm but now knowing the bearing to return to Luepa and Uruyén.

We had arrived unannounced, but we were soon surrounded by frontier guards, priests, and Taurepán Indians. There was then no regular transport to that spot from anywhere, but it is near the Brazilian border and such contact as there was with the outer world was then almost entirely on foot or horse by trail from Brazil. Hence the small guard unit, and hence the presence of horses and cattle, beasts then still unknown elsewhere in southeast Venezuela. In the motley cavalcade that galloped to investigate our strange apparition was a cordial priest from the mission who had cleverly coped with his beard by tying the ends together on top of his head. We were soon mounted, too, and grandly escorted not to the mission but to the house of Lucas Fernández Peña, earliest non-Indian settler in the Gran Sabana. The exceptional nature of that establishment was evident even from afar by such details as the long strings of beef hung out to dry, producing the luxury of charqui.

We had somewhat regretted leaving the Uruyén camp for what we expected to be a hut in Santa Elena, and the Peña home was a great surprise, more comfortable than any place we had been in Venezuela outside of Caracas. We were indeed in what was then still one of the world's most remote and inaccessible places, but no one would have imagined so as we sat in chairs and ate a good dinner off a table in an attractive room with a charming host and hostess. The hostess was not the Sra. Peña, an Indian who preferred to tend her cooking outside, but the oldest daughter, a beautiful girl with the best features of both races. (When we came to leave, I gave her my folding cot as a hostess present, and my wife, who is a psychologist, professed to see something Freudian in this gesture.)

One's perspective changes somewhat after camping for months in Venezuelan backwoods and wilds. The floors of the Peña mansion were in fact dirt, the palm-leaf roof sheltered bats, and tarantulas scrambled around the walls as we ate, but luxury is relative and this was relative luxury. After dinner, we set up our cots in a large, clean room, white-washed and nicely decorated with blue stenciled patterns, and grate-fully, finally relaxed.

I had had and was to have longer, more difficult, even more exciting days, but for 30 March 1939 that was enough. It will do here as a sample, or an appetizer.

2
Meanwhile Back at the Family Tree

Where did you come from, baby dear? — George Macdonald

It is alleged that a Frenchman of humble origin, when asked from whom he descended, replied, "I am an ancestor, not a descendant!" That was later borrowed by Don Marquis and has passed into folklore as the perfect squelch to a snob. I agree with that sense of values, but I do think it of some use to consider anyone's descent. If he has done better than his ancestors, his accomplishment is all the greater; if he has done worse, he at least has something to compensate the failure. In either case, our ancestry does help to explain us, and may sometimes even help to excuse us.

The Simpsons originated as a cadet line, sons of a younger son Simon, of the Scottish clan Fraser. In true Scottish tradition they seem to have been careful men with pennies, and Simpson became the typical pawnbroker's name in Britain. Our branch apparently moved to England long ago, but seems not to have done anything as profitable as pawnbroking. There is a legend about our ancestors around the time of the Restoration. It was told to me by someone who was more often mad at than about the Simpson side of the family and so the story is probably not true, but I hope it is. According to my informant, the Simpsons long cherished an elegant ancestral costume. They assumed it to have been court dress, but any more precise tradition had died out. They therefore took it to a historical museum to determine what rank in the peerage the wearer would have had. The curator informed them, so the tale goes, that the costume was a flunky's livery.

I have not seen a Simpson genealogy, although I believe one exists. The earliest Simpson ancestor whose name I know is William A. Simp-

9

son, my great-grandfather, who lived in Boston early in the nineteenth century. My unreliable informant once told me that William's father was a British sea captain trading to Canada and that the family settled in Boston as storekeepers. All I know for sure is that I have seen a certificate granting relief to William A. Simpson as a distressed simple sailor born in Massachusetts. In any case, from Boston William moved to Chicago and there my grandfather George Washington Simpson spent his whole life except for ninety days marching through Georgia as a buck-private Illinois volunteer in the Union army. He was the only grandparent I ever knew, and I saw little of him. Except for being a man of slight stature he looked extraordinarily like George Washington, a resemblance he doubtless cultivated. I am named for him, hence at one remove for George Washington, but I do not at all resemble either of them. As I remember my grandfather he was a quiet and kindly old man, usually bedridden from asthma and other ills, most interesting to me, then a lad, because he used to play a Spanish guitar while propped up in bed.

George W. Simpson married the daughter of a Welsh Presbyterian minister, the widow of another Welshman. By her he had three children, one of whom was my father, Joseph Alexander Simpson. My father's half brother, fruit of my grandmother's first marriage, was noteworthy not only for bearing the name John Jones but also for being our nearest relative ever to accumulate any considerable sum of money — not that any Simpson was ever much benefited thereby.

The ancestry of my maternal grandmother, born Sarah Capen Dimond in 1842, is richly recorded and well known to me because my mother took much verbalized pride in it. It includes a colonial governor of Connecticut, but even more especially one Thomas Dimond, who settled in Fairfield, Connecticut, sometime in the 1630s and died there in 1658. In the freewheeling way of those days his name was also spelled Demont, Demond, Deming, and Dymont, and several of his descendants spelled it Dimon. My great-grandfather Henry reverted to the spelling Dimond, and it is as Dimonds by another name, indeed more or less six others — that my sisters and my children are registered as Hawaiian Mission Cousins, an organization of descendants of missionaries to the islands.

Henry volunteered as a missionary to what were then more often called the Sandwich Islands, and since unmarried missionaries were not accepted, a marriage was arranged with one Anne Maria Anner, the daughter of a New York jeweler. There is a romantic and almost certainly untrue legend about one of her ancestors, too. According to the story, on a fateful eve of the French revolution a richly dressed boy was smuggled on board a ship in a French harbor. The captain was entreated to save the boy, whose name and rank could not safely be revealed and who would be killed if the revolutionaries discovered him. The boy was saved and lived thereafter under the captain's name, Anner. The implication is, of course, that as I am the senior living male descendant of that boy, the French nation should by rights be taking orders from me.

Anne Maria Dimond née Anner kept a journal of the horrible trip from New York around the Horn to the Sandwich Islands. You can read it almost word for word in Michener's novel *Hawaii*. Her marriage began as one of convenience but evidently became one of affection, as the union was long, happy, and fruitful. Anne Maria and Henry died before I was born (they would both have been ninety-eight in the year of my birth), but I feel that I know them well. They brought up my mother, and especially in her last years she talked lengthily about them. She was far from being an uncritical woman, but almost all her accounts of her grandparents were affectionate. She insisted that I closely resemble my great-grandfather Henry Dimond, not physically but in temperament. She obviously was mistaken about that however, because she also said that Henry's Hawaiian nickname was *pa'akiko*, which means approximately "pig-headed."

In view of the mass of available information about the ancestry of my maternal grandmother, most of which I have spared you, it is odd that practically nothing was known to my mother or is known to me about the ancestry of my maternal grandfather. He was J. Russelle Kinney, probably of at least partly Irish descent. Kinney is said to be an Irish patronymic, perhaps not a very distinguished one as I find only a few not particularly famous people of that name in reference works. Russelle (accent on the -*ell*) studied medicine in Vienna and after marrying Sarah Dimond, daughter of Anne Maria and Henry, in Honolulu he moved with her to Cedar Rapids, where he was company physician for a railroad. A number of Sarah's letters survive and they make grim reading. The life of a company doctor — or I suppose of anyone — in a small town back in 1860s and 70s was not attractive by our present, perhaps degenerate, standards. Anyway, poor Sarah shortly died of it. Russelle shipped the babies back to their grandparents in Honolulu and took off. My mother did not know where or when he died but had heard that he married again and became a drug addict, not necessarily in that order or as cause and effect. In fact drug addiction was long an occupational hazard for overworked doctors. This had a curious effect on my mother, curious because she had no contact at all with her father's addiction if, indeed, it really occurred. She had a lifelong fear not only of addictive drugs but of almost all medicines.

Like many mission children, in due course my mother was sent to Oberlin College. I am uncertain as to just where and when she met my father, then a Chicagoan in his twenties, but they were married in 1891. In 1895 they produced a daughter, Margaret Anna, and in 1898 another, Martha Helen. At 9:15 A.M. central standard time on 16 June 1902 in an apartment house on the south side of the Midway, a son was born, and in due course he was christened George Gaylord. My father had started to practice law in Chicago and the Gaylord was for Truman Gaylord, another lawyer with whom he was in brief association. (I do not remember ever meeting him or hearing from him.) Mother, who (to put it mildly) was not entranced by her in-laws, avoided calling me by my Simpson grandfather's name by always addressing me as George Gaylord in full. The long-drawn-out cry of "George Gay-ay-ay-lord" of an evening in attempts to bring me to supper or to bed became a jest for my contemporaries. My father called me "Jig" when I was young and switched to simple "George" as I acquired dignity in his eyes. I was George to others even as a child and still am to most. After I was adult, however, my sisters and my childhood friend, later my second wife, Anne Roe, started calling me "G", and that is used by a few others, mostly friends who came to know me through or with my wife Anne. One or two brash souls have thought it especially chummy to call me "Gee-gee," which I find offensive.

I have hinted that mother did not get on too well with my father's family. As regards her mother-in-law this became an almost obsessive hatred, which burned brightly to the day of mother's death, a half century or so after my father's mother had been summoned to her reward, presumably in the heaven of Presbyterians. This feeling probably had some bearing on my father's moving the family away from Chicago and the Simpson-Jones ambit. We went briefly to Cheyenne, Wyoming, and then while I was still an infant settled in Denver, Colorado, which I did not leave definitively until I was twenty years old. My earliest memories, from the age of three, date from well after our move to Denver. One is of an aunt, one of my father's full sisters, giving me a set of lead soldiers. It seems out of character for me to remember just that; probably it is because never again did any of my several aunts on both sides give me anything but trouble and bad advice. My other earliest memory is that after giving the matter much thought I decided that our street was called Vine because our house had vines over the porch and that Sunday was so called because it rains less often then than on weekdays. It was, I believe, in character that I was already hunting for explanations; I am not sure how typical it was that both my hypotheses were wrong.

3
Little or Much?

A little learning is a dangerous thing. — Alexander Pope

Much learning doth make thee mad. — Saint Luke, if he wrote the Acts of the Apostles

Learning is my trade, and I consider it both interesting and important, therefore an appropriate topic here. However, my personal formal exposure to learning can be only moderately interesting and only secondarily important, if at all, to anyone but me. It is here outlined for the record, in sequence, briefly, and skipably.

In two years of kindergarten, 1906-08, I learned nothing except that you must not steal modeling clay even if you covet it and do not know any other way to get it.

In first grade I learned to read and write; I was not one of those geniuses who can read English at two and Sanskrit at four. I do not recall that I learned anything in the next five years of school (I was out of school a year, but also skipped several grades), although I learned a great deal out of school. I was then vaulted over eighth grade, the last of grammar school (there were no junior highs) into —

East Denver Latin School, which long since ceased to exist either in name or in practice. It included only ninth grade. I was threatened with being bounced back into grammar school because the teacher couldn't make me see what algebra is about. I did suddenly see it one day in spite of the class, and after that through high school and college never got a grade lower than A+ or 98% in a sequence of mathematics courses. A college professor later urged me to become a professional mathematician but I explained to him in my kindly teenage way that I couldn't do that because unapplied mathematics isn't *about* anything.

While at East Latin I also consumed a great many cream puffs at five
cents each from a street vendor during lunch hours.

For three years, with some more time out for illness, I went to East
Denver High School, which still exists in name only, being in different
buildings in a different part of town with different kinds of teachers and
pupils. There I learned to recite Coleridge's *Ancient Mariner* from beginning to end, which would be a great social asset if I could still do it, but
I can't. I also had my first course in science and was meant to learn that
science consists of measuring various things in different ways and making the answers agree with the book. That calls for another epigraph
even here in the midst of a chapter.

*It is the extreme of madness to learn what must then be
unlearned.* — Erasmus

While at East Denver High I also rode no-hands down Capitol Hill in
snow storms on a bicycle sometimes named Borak and sometimes
Bucephalus.

I was graduated in a mass ceremony for all five Denver high schools
in the Denver Civic Auditorium shortly before my sixteenth birthday.
That was 1918, in the final phases of the Kaiser's war, and as the bright
and shining faces advanced in cohorts to receive their diplomas they
marched under a banner with the stirring words *"THEY SHALL NOT
PASS."*

In fact, we had managed to pass in Denver as the Germans failed at
Verdun. In 1918 I entered the University of Colorado and, for a first
tussle with the army, the Students Army Training Corps. Between the
army and the great flu epidemic, all I learned as a college freshman was
that one should never volunteer. Even that was not well learned, for I
did again volunteer for the army in Hitler's war.

In 1919-20 I couldn't financially manage school and was out for a
year, doing a great variety of jobs in Chicago, on the road south from
Chicago, in Texas, and in the Colorado mountains. In the autumn of
1920 I could return to school, assisted by a scholarship for tuition, fur-
nace-tending and snow-shoveling for a room, and table-waiting and
dish-drying for food. And at long last I began to learn a few things that
did not soon or completely have to be unlearned again.

I went to college thinking vaguely that I might become a writer, but
I was soon bothered by the fact that unapplied writing, like unapplied
mathematics, isn't *about* anything, anything real, that is. (I do not have
this objection to reading; that can well be about things that are not real,
or particularly that transmogrify the real.) To write, I wanted something both real and worth writing about, and this feeling, which started

for me then, in 1920, has continued to develop for well over half a century so far. It started with a course in historical geology, not that I then yet had either the good sense or the self-confidence to think of myself as writing anything worthwhile on so exalted a subject.

My first teacher in what became my field, in a broad sense, was Arthur Jerrold Tieje, and if I thought that such tokens of appreciation reached their object I would ever since have kept a candle burning for him. He was unpopular with almost all the other students because he expected them to work hard and long, but I liked to work hard and long. He was excited about historical geology and paleontology. He made me see why they are exciting, and that made me excited about them too. He left Colorado in 1922, went to the University of Southern California, and died at the early age of fifty-three in 1944.

When he decided to leave Colorado he urged me and my parents that I transfer to Yale as the best place to lay a foundation for a career in paleontology. The family finances had by then improved a bit, we decided that it could be managed, and it was. So I left Colorado. In 1968, just fifty years after I had matriculated there, I returned, and while an honorary doctor's hood was hung on me I was introduced as the man who had taken longer than anyone in history to obtain a degree from that university.

While at the University of Colorado I was also disappointed in love, or so I thought at the time, and I purchased a fair number of hamburgers, with everything to go, at a corner stall in the late evenings or early mornings, at ten cents apiece, little as I could afford them.

The curriculum of Yale College was considerably more demanding than that of the College of Liberal Arts of the University of Colorado, and I spent my senior year in a sophomore dormitory studying subjects such as history, economics, and languages that I had thankfully been able to dodge thitherto. I was still short a year of French at commencement and took off for France, where I spent the summer living (cheaply!) with a family that spoke no English. On my return I passed an examination in second-year French, and was considered to have been graduated with the class of 1923, although I had never met any of my "classmates." In any case, I was able to continue at once in the Yale graduate school, which was the intention back of the transfer.

It was seventeen years from my entering kindergarten to my entering graduate school. Even granting that two years of analphabetic kindergarten were not particularly academic and that I had been out of school three years in all from illness, poverty, or both, that is a long time to sit in classes and to have rather little to show for it. I had at least been growing up, although not yet quite enough. I had learned a great deal outside of classes and had experienced many interesting things, although on both accounts not yet even remotely enough. But in graduate school for the first time I felt, and still do feel, that I was getting a full hour's worth for each hour of schooling. I was fortunate in my fellow graduate students. Students often learn more from each other than from teachers, and by great fortune I happened into geology at Yale when most of the students were exceptional. Not long thereafter they could count among their ranks a surprisingly large percentage of eminent American geologists.

The faculty of that time was also exceptional. I had fixed on vertebrate paleontology as a specialty, and Richard Swann Lull became my major professor. Other faculty paleontologists were Charles Schuchert, retired but decidedly there and active, and Carl O. Dunbar for invertebrates, and the eccentric but enthusiastic paleobotanist George R. Wieland. Paleontology bridges geology and biology, and the graduate zoology department was also then strong and inspiring at Yale. Work with Ross G. Harrison, an important figure in the history of biology, was particularly helpful to me.

One of the major assets of Yale paleontology is the Marsh collection. Supported by his uncle, the New England banker George Peabody, Othniel Charles Marsh (1831-99) had amassed a great collection of fossils of dinosaurs, ancient mammals, and other vertebrates, and had built the Yale Peabody Museum to contain it. When I went to Yale the old museum had been torn down and the collections were scattered almost inaccessibly around the campus, but while I was there the present Peabody Museum was completed and the collections united and installed in it. Professor Lull, age fifty-five when I became his student, had long ago given up field work on the grounds that the best fossil collecting he knew was in the basement of the Peabody Museum. A devotee of vigorous outdoor life myself, I considered that specious reasoning, but I profited by it. I went collecting in the basement and decided that the most important fossils there were the exceedingly rare remains of the earliest mammals, those of the Mesozoic era, the so-called Age of Reptiles. Professor Lull agreed so thoroughly with that evaluation that he considered them far too good for a brash graduate student. However, he either forgot that ban or changed his mind, for he did let me go ahead when I started working on them. I produced a dissertation entitled "American Mesozoic Mammalia." In June 1926, I attended my first graduation ceremony since 1918 and was given that academic license, an earned Ph.D. degree. So ended my formal schooling.

Learning, both little and much, by then must have made me both dangerous and mad if Pope and Luke are trustworthy.

While at Yale I was married and became the father of two delightful daughters.

Note

§ Like many academic persons, I am a footnote addict. One of my favorite books is the Yule-Cordier three-volume edition of Marco Polo's travels, which consists almost entirely of footnotes with Ser Marco's text lightly and obscurely sprinkled among them. I have never reached that level of grandeur, but some of my early publications are so liberally footnoted that a colleague once remarked that we seemed to be corresponding in published footnotes rather than by letters. As befits advancing age, I have become somewhat more abstemious in that respect, if in few others. Nevertheless I do intend to footnote a bit in this work, but will keep the pages clean by appending the notes to the ends of chapters rather than putting them at the bottoms of the relevant pages, where I personally prefer them.

Just now, I only want to assign a little collateral reading:

References

Clements, T. D. 1944. Arthur Jerrold Tieje (1891-1944). *Bulletin of the American Association of Petroleum Geologists* 28: 686-87.

Schuchert, C., and C. M. LeVene, 1940. *O. C. Marsh: Pioneer in Paleontology* New Haven: Yale University Press.

Simpson, G. G. 1929. American Mesozoic mammalia. *Memoirs of the Peabody Museum of Yale University*, vol. 3, pt. 1, pp. i-xvi, 1-171.

———. 1958. Memorial to Richard Swann Lull (1867-1957). *Proceedings of the Geological Society of America for 1957,* pp. 127-134.

4

My Careers as a Musician

Most of the rest of this book will be about adult matters, but there are two subjects I want to include that start in childhood. One is light-hearted, indeed frivolous: this chapter on my careers as a musician. The next is very serious indeed and its title, "God and I," is not meant to be taken lightly. So, first music, then God. In order to relate the two, I give here some slightly delayed epigraphs:

> There is no truer truth . . . than comes of music. — R. Browning.

> Music tells no truths. — P. J. Bailey

> The God of music dwelleth out of doors. — E. M. Thomas

> There is no music in nature. — N. R. Haweis

> Music religion's heats inspires. — J. Addison

> He was a fiddler, and consequently a rogue. — J. Swift

And finally, most immediately relevant:

> Hell is full of musical amateurs. — G. B. Shaw

In the Denver neighborhood where I grew up nearly all the kids went through the tribal rite of taking piano lessons. That was one of the Advantages — along with Christian indoctrination and tooth straight-ening — that our parents felt they owed us. My teacher was a sere spin-ster, possibly as old as thirty-something since I remember her as elderly. She prided herself on being a Good Fellow and was known to parents as one who Understood Children. Of course her juvenile clients, the boys, at least, disliked her. Having her put a comradely arm around one's shoulders was embarassing as well as unpleasant.

We felt a certain awe, not mixed with admiration, for the girls and sissies who rose to the dizzy height of playing pieces with crossed hands. A few, including Anne Roe, whom I later married, actually achieved a workmanlike mastery of compositions not written down for the juvenile trade. Later, in high school and college, this accomplishment was really valued. It provided accompaniment for singing and, more rarely, dancing. Most of us never reached such heights. My own progress stopped when I could play two or three notes almost simultaneously with each hand, uncrossed. It was impossible to trick my nerves and muscles into any more intricate coordination, and there my career as a pianist came to a dead stop.

Perhaps the fault was lack of concentration, for I was working as a coorganist at the same time as my piano study. The duty of the organist at the Capitol Heights Presbyterian Church was merely to push the keys and pedals as indicated. The coorganist had less glory but more responsibility. He crouched in a small cubicle beside the organ's works, hidden by a cloth curtain. There he kept his eyes focused on a nail tied to the end of a string. When the nail rose to a certain height, that indicated that air was getting low, and then the coorganist seized a long wooden lever and pumped it up and down vigorously until the nail fell and the sound rose. Such, at least, was the theory. There was, just barely, enough light to read in the cubicle and I was usually so engrossed in some highly nonreligious book that I did not notice the rise of the nail until the sound was about to expire with a final squeak. Hymns played with me as the coorganist had exaggerated diminuendo and crescendo unintended by the composer but entrancing to all.

The pipes in this organ were set loosely into holes in the tops of wooden boxes. Over considerable ranges the holes were the same size even though the pipes produced different notes. I discovered, in the course of my post-keyboard study of the organ, that it was possible, indeed irresistibly necessary, to reassort the pipes in such a way that their relationship to the keys was erratic and unpredictable. This principle mastered, I prepared for the climax of my career with the organ by thoroughly scrambling the works one Sunday between Sunday school and church. The congregation assembled, quieted, and rose to sing the Doxology. My colleague, the organist, raised her hands high in the air and brought them crashing down for the first loud "Praise." The result was a most appalling discord that must have sent all the dogs waiting outside into ecstasies. The deafer members of the congregation went quaveringly on through ". . . God from whom all blessings flow." The organist, baffled but not yet beaten, raised her hands and looked at her fingers in a puzzled way, while the less deaf part of the congregation stood with silent mouths open. A second start only duplicated the first. Even the deaf now sensed confusion in the air and wailed to a halt. By then I was off and running. They caught me, but it was worth it.

One result of that and other exploits in the same spirit was a much-needed reoganization of the Sunday school. A half dozen of the most energetic boys were made into a special class, meeting out of sight and sound in the church tower in charge of the most muscular of the adult male faithful. This powerful Christian could lick all of us put together, and once that fact was fully established everything went smoothly. The good doctor (he was the neighborhood M.D.) had read about casting pearls before swine, so he never bothered with the allotted lessons. Instead he told us about his adventures in South America, and to this day I associate the biblical hell with boa constrictors and alligators. (Now that I know something of South America at first hand I realize that practically nothing the good doctor told us was true. I admire him all the more, and his teaching was probably as true as that in the other classes, besides being far more entertaining.)

But my career as a Christian is a different subject. Let us return to my careers as a musician.

After it was evident that I would never be able to cross my hands on the piano keyboard, I took up the flute. That instrument has the advantages both of not requiring crossed hands and only producing one note at a time, so it seemed reasonable to expect that even I could master it. In my father's day the flute had been a favorite courting instrument. He had not merely one but two. I never asked what this implied about his courting activities. Both were wooden instruments with open holes, a system even then long obsolete among professional flautists. I was given the older of these two antique instruments and sent out on the prairie to learn to play it.

Being sent out on the prairie was not merely to put me beyond the range of suffering ears. The prairie began at the end of our block and a mile or so farther there was a cabin occupied by a plump, unshaven German who claimed to have had a distinguished career as a flautist. Now he had no visible means of support; as far as I ever learned I was his first and last pupil. At any rate, his own tootling sounded fine to me, and he had enough respect for the instrument to be lacerated by my efforts. For preparation, endurance, and compensation he drank forti-fied California wine from a gallon jug before, during, and after the lesson. When I knocked on his plank door he reached for the jug without a word of greeting, but he was almost cheerful after the lesson when I placed a silver dollar in his hand and trudged away across the prairie.

By the time I went away to college I could handle almost any tune on the flute, provided, of course, that the tune was simple, slow, and in the key of C. During my first semester in college (this was in the fall of 1918) I was in the Students Army Training Corps, and I volunteered to play in the S.A.T.C. band. There were a few hardened army men who had already been in the corps for two or three months, and they warned us never to volunteer for anything. Nevertheless, volunteering for the band was obviously a cinch. While the others learned the manual of arms and close order drill in the slush of the parade ground, we bandsmen would be sitting dry and warm in a practice room. To qualify for the band you had only to produce your own instrument and play a scale on it. I quickly sent home for my flute and was in. The one volunteer who could sing a scale but did not have an instrument obviously had to be bandmaster. The day after we were organized we learned that the old army men were right again: band practice was extra duty and did not excuse us from regular drill.

Fortunately we had no band arrangements with a flute part. I played any wind part roughly within the range of my instrument, and so had no solos or other passages in which my flute could really be heard above the general uproar. After the first bar or two I was always lost and so carried on twiddling my fingers at a great rate but careful not to make any sound. When the end was approaching, I looked ahead to spot my last note and got carefully set for it. I hit that note loud and true, and after the others stopped I held it long enough to be sure the bandmaster knew I had been right in there fighting. Our only public performance was on Armistice Day, when the band proudly led the whole corps on parade through the streets of Boulder, Colorado. We were a little downcast that we had never been under fire — not even from our comrades, for we had never been issued ammunition for our rifles. Yet we were also excited and exalted by the occasion, and never have last notes been played louder or longer on a flute.

Drama would demand that my career as a flautist end there. In reality it tapered off so quietly that I cannot remember when it stopped altogether. For two or three years I played in Sunday-school orchestras. There were several hymns in which I could play the soprano melody all the way through. It is not true, as some of the music critics have it, that I was exclusively a last-note player.

My most enduring fling with contact, as opposed to spectator, music has been with the mandolin. My sister Marty, an artist, had somewhere acquired a picturesque but cracked mandolin as material for still lives. Between paintings she picked out tunes, which so fascinated me that I got a mandolin of my own. I never had a lesson in my life, but within a year or two my friends were hardly laughing at all when I sat down to the mandolin. I could play several scales and was working my way gingerly into Mexican folk songs and Gilbert and Sullivan (all but the fast ones or those with more than one flat).

Marty moved to California, Hitler's war came along, and the mandolin was forgotten. After the war it was dusted off again for the Mandolin and Martini Club. This select organization of gin-, music-, and whiskey-lovers arose from some subtle sort of spontaneous combustion on an unrecorded and vaguely remembered date. My second wife, Anne Roe, played the guitar and piano, a doctor played the banjo-mandolin, a publisher the first mandolin, and I the second mandolin. "First" and "second" here refer to skill; actually the second mandolin played the solo or melody parts because they are (as we played them) easier. The publisher's wife and the doctor's nurse were the club's official cupbearers and generally joined in raising voices in song toward the end of a hard session. We occasionally had guests, but for some reason no guest ever returned for a second meeting. The specialty of the club was American folk songs. We had dozens of those in our repertory, but we always wound up with the "Hatikvah," the national anthem of Israel — a rousing finale regardless of nationality or ideology.

Like all things good or bad, that came to an end. First the doctor, who had been dividing his practice between New York City and upstate New York, eliminated the city part. Then the publisher moved to a prosperous suburb. Finally I went to Cambridge, and in New York there were none. Still there are rumors that when the moon is full and the cups are too, a ghostly tinkling as of mandolins can be heard from a closed room on East Holmes Street in Tucson.

References

Bickford, Z. M. 1920. *The Bickford mandolin method: Book I.* New York: Carl Fischer.

Boni, M. B., ed., 1947. *Fireside book of folk songs.* New York: Simon & Schuster.

Harris, H. J. 1950. *Brucellosis (undulant fever) clinical and subclinical.* 2nd ed., revised and enlarged. New York: Harper & Row. This has nothing to do with music and there are no mistakes in it, but if there were it would be the fault of the Mandolin and Martini Club.

Reid, J. M. 1969. *An adventure in textbooks.* New York and London: Bowker. I consider it typical of a publisher that although I drew his attention to the error in proof on page 114, where he calls the Mandolin and Martini Club "The Mandolin and Whiskey Club", the author refused to make the correction.

Smith, W. J., presumed editor. 1953. *100 solos for mandolin or violin . . . also adaptable for autoharp, pop piano, accordian, banjo, etc.* New York: Wm. J. Smith Music Co.

5

God and I

No poet or seer has ever contemplated wonders as deep as those revealed to the scientist. Few can be so dull as not to react to our material *knowledge of this world with a sense of awe that merits designation as religious.* — G. G. Simpson

This no Saint preaches and no Church rules. — Matthew Arnold

I have heard that there is a physicist who lectures on "The Universe and Other Things." To write about "God and I" may seem equally ambitious and even more arrogant. It would be better if it could be written "god and i." (I have also brooded on "God and i" and even "god and I," but neither is exactly right.)

It would never occur to a child that either ambition or arrogance could be involved here; at least it could not have occurred to me. We were what is curiously called a God-fearing family, which meant that we had been persuaded by preachers, who cannot have understood the Bible, that Jehovah is lovable, not fearsome. My good friend the milkman Tom McGovern, suspected of being Catholic through no fault of his own, and a Polish family, Jewish like Jesus, were exceptions, but almost all my boyhood acquaintances attended Capitol Heights Presbyterian Church. We were perhaps more attentive than most, for we commonly attended three services on Sundays as well as the midweek prayer meeting, and in addition had family prayers and psalm recitations. At the age of nine I formally joined the church, having assured the minister that I believed the Presbyterian creed, although I think the matter of infant damnation and some of the other points in fine print were not brought to my attention.

When I was nine the verb "to believe" was just beginning to have some meaning for me. It has little or none for the very young. The question of belief or disbelief does not then really arise in connection with Santa Claus, Christ, the Land of Oz, or heaven. How many children ever asked themselves, "Exactly what real, objective evidence do I have for this? Is it sufficient to make the existence of God, the Blue Fairy, or the Grand Lama of Tibet more likely than not?" What young child ever realized that reading something in a book or hearing it from a pulpit is not necessarily evidence? What young child ever cared, or ever so much as thought of these matters in that sort of frame of reference? That is what is known as "childlike faith," which is not really faith *in* something in the positive sense of rational acceptance, but merely ignorance of the processes of rationality. It is not unbecoming, is indeed fitting and may have an attraction in the inchoate, growing spirit of a child. It is most unbecoming, indeed repulsive in a nominal adult, and those who exhort grown men and women to have childlike faith are enemies of the fully human spirit.

Moments of conversion have been precisely indicated, notably that of Saul on the road to Damascus. Again I am putting myself in company where I do not belong, but I can indicate my moment of deconversion. Presbyterians are Christians. (In our milieu one could almost have posited, "Christians are Presbyterians.") I was a Presbyterian. Therefore I was a Christian. The logic is impeccable, but as I grew only a little older and vaguely more thoughtful I began to sense the possibility of fallacies.

> *George:* When do you really become a Christian?
> *Preacher:* When you are born again.
> *George:* What does that mean?
> *Preacher:* When you don't want to be naughty any more.

Soon there came a lovely morning when I awoke feeling holy. I was reborn! All day I went my Christian way, but that evening the moment came. I had not been naughty; I had no immediate plan to be naughty; but I certainly had not lost all desire to be naughty as occasion might arise. I had not been reborn. I was not a Christian. As to what, in fact, I was, I did not know.

The incident is trivial. One does not turn from a Presbyterian to an atheist at a moment of deconversion, and indeed I never became an

atheist. But it is a fact that I never again considered myself a Christian
and that I then started the long, always difficult and sometimes desperate process of unlearning all that had to be unlearned and learning as
much as I could of what had to be learned in order to find out what I
was and am.

Thus the process started with a rather silly discrepancy between
dogma and reality. I became well aware of other discrepancies. It is too
obvious for comment that Sunday saints are commonly weekday sin-
ners. Preachers do not practice all they preach. The doctrines of Christ,
as far as they have come down to us, are in some fundamental respects
incompatible with twentieth-century American mores, and then they
are quietly ignored by the nominal Christians. The Bible contains at
least as much hate and horror as love, and in any case the concept of
Christian love is rarely considered applicable to those who do not
believe exactly as we do.

I observed those and the many other shortcomings of the actual practice of Christianity with interest but not with dismay. It was not they
that led me even farther from dogmatic religion and toward religion
that could be maintained without sacrifices of rational and decent
humanity. There are several reasons why I am not here going to relate
the full course and details of that long process. It was indeed such a
long process that it would require a book in itself, a book quite different from this. Moreover, I am not in any sense evangelical. I am not
intent on persuasion. The search is valueless unless one makes it for
oneself, although an account of the basis for another's search and its
outcome can be of interest if not assistance. That, briefly, is all I shall
attempt before leaving the subject for good as far as this book is
concerned.

First, I might just mention that formally leaving the church proved to
be harder than joining it. The mere fact that I rejected some beliefs on
which church membership was predicated was not accepted as a reason
for striking my name from the rolls. Once a Presbyterian always a
Presbyterian, if only to maintain census figures. Membership in one
congregation could be ended only by transfer to another, which was
accomplished by a pastoral "letter." On grounds that I was leaving
Denver for a time, I obtained my "letter," then burned it instead of
handing it in to another minister.

Later, when we had young daughters in New York City, my wife
Anne and I joined a Unitarian church, the only organized sect to which
we could subscribe without hypocrisy. Our reasons were partly social;

and we believed that learning something about various formal religions helps young people to understand the culture in which they are to live. When we left New York and the grown children had left home, we also left the Unitarian church, in which continuation of membership is not compulsory. Our leaving the church was not based on any disagreement, but only on the feeling that there was no real advantage, to us or to others, in continuing to connect our personal religion with an organization and a ritual.

It is worthy of remark that my parents, earnest and even bigoted Presbyterians when I was a child, also became non-Christian Unitarians in late middle age and so continued to their deaths. Their deconversion occurred after I had left home and without discussion of religion between us. It resulted from their own capacity for learning and for spiritual growth at ages when most people have lost both.

I am a scientist, and my intellectual and spiritual motivation involved the development of a basically scientific view of life and the world. That can be misunderstood. The scientific attitude, as here considered, is not confined to scientists. In fact it is fairly — though not sufficiently! — widespread and is inherent in what is rational in modern civilization. It is not shared by fundamentalist preachers, astrologers, flower children, most generals, or many bureaucrats, but fortunately none of those classes has yet quite taken over the country or the earth. The scientific attitude has nothing whatever to do with manufacturing cars or dropping bombs, and very little to do with inventing them.

The impulse of the scientific attitude is the desire to learn, to learn about ourselves, about others, about the earth, about the universe. Its criterion is that what is learned should be real, should in fact exist in ourselves and onward into the universe. It is, then, necessary to have a concept of reality and to make judgments of evidence as to what is real and what is not. To put myself again in company above myself, I might paraphrase that when the wind is southerly one must know a fact from a handsaw. If a preacher says, "We were created in the image of God," the fact is that the preacher said it, and that is the only fact established. Whether in fact a god created us in his image requires a judgment of evidence not adduced by the fact that someone says so.

Oversimplifying and cutting some philosophical corners, one can say that a fact is something that can be observed and that can be confirmed by the observations of others. Worthy belief or, in some sense, truth is interpretation of fact, and part of the scientific attitude is that any such interpretation is subject to correction. Truth in this sense can only be relative or tentative. That is hard for any of us to take, intolerable for children, young or old, who cannot tolerate uncertainty. One can, at least, be relatively certain that a belief contrary to the weight of factual evidence is not true. In the example previously given, there are indeed many known facts bearing on the origin of man, and these facts are strongly inconsistent with the creation of man just so, in his present form, and hence in the image of any given being or Being. Therefore belief in that statement is not respectable. To say that is not to deny God or god. It is to require a reevaluation or redefinition not of the facts but of the statement.

This is only an example of the basis of an enquiry that prefers reality to illusion and evidence to superstition. It also illustrates the approach to an adult religious reorientation that is not a denial but an affirmation. The long, magnificent story of man's evolution from the stuff of stars is incomparably more wonderful, more awe-inspiring than that fable of an anthropomorph making mudpies in a mythical Garden of Eden. Further, the emotions of love, beauty, hope, sympathy, or, darkly, of hate, ugliness, despair, and scorn are real and are inspiring and enjoyable or degrading and painful no less because they are material facts with material explanations.

Although some scientists neglect this, it is a necessary axiom of science that no supernatural postulate be proposed for interpretation of material (which means observable) phenomena. We are far from being able to explain all known material phenomena, but we cannot rationally hope to proceed toward explaining them in any way but by naturalistic interpretation of observed facts.

But that finally brings us up against a mysterious ultimate: The Mysterious Ultimate. Much has been explained and much more is certainly explicable in terms of the material characteristics or properties of the universe. That does not explain how it happens that the universe has those characteristics or properties, how, indeed, it happens that the universe exists. I, at least, cannot even imagine any possible facts, any conceivable observations that would lead toward such an explanation. That Mysterious Ultimate is, then, inaccessible to scientific, which is to say to rational, human investigation. As far as I am concerned, it is God, or better, god.

This god is in full literality ineffable, which means incapable of being expressed, unutterable, indescribable. The evangelistic mountebanks who speak of their god as ineffable and then proceed to describe him, even as if he were a rather odd but powerful human to whose wishes or commands they are privy — those characters are not only mountebanks but also fools, crooks, or both. To speak of God as an intelligence not only goes beyond knowledge but also is silly at best, blasphemous at worst. Intelligence is a limiting characteristic, material and specifically animal. Whatever any god may be, surely he (or it) is neither material nor animal. To think of God as personally loving us is silly or blasphemous or both for the same reason, and moreover is obviously contrary to the weight of all available evidence — and even the evidence of the Bible if that were accepted.

Perhaps I have made it all too clear that I have no patience with dogmatic religions. I should also make it clear that I find them extremely important, interesting, and worthy of study. Although not quite universal, religion is one of the most widespread traits of the human species, and it has usually taken a dogmatic form even in relatively primitive societies. By "dogmatic" I mean having defined tenets to which the accepted communicant or group member must subscribe without or even contrary to factual evidence available to him. That is manifestly true of all the great organized world religions: Christian, Moslem, Hindu, Confucian, Buddhist, Shinto, Taoist, Judaic, and Zoroastrian. It is also true of the many small tribal religions, with less or in some instances virtually no organization.

It can be no accident that dogmatic religion became nearly universal in humans and, as appears probable from available evidence, at quite an early stage in the evolution of *Homo sapiens*. The general reason seems to have been that it was adaptive in both a biological and a sociological sense. At the tribal level and for an as yet undefined distance beyond, dogmatic religion tended on the whole to promote the cohesion, empathy, welfare, and so the survival and growth of societal units. It is not even an apparent contradiction that such religion also tended, and still tends, to some brutal persecution and warfare between such units.

Perhaps the most interesting aspect of this is that the truth of dogmatic religion has had no bearing whatever on its adaptive value. So much is obvious, if only from the fact that all the major dogmatic religions surviving today have dogmas such that no two of them can both be true.

There is a good deal of evidence that dogmatic religions, almost certainly adaptive among tribal or primitive peoples, are now becoming inadaptive. I believe that to be one of the crucial problems today, but it would be cumbersome and digressive to pursue the topic further here. I can close the chapter, with some risk of smugness, by the statement that I am at peace with my god and that I would not care to spend eternity with the Presbyterian God (or some of His devotees).

Note

§ Almost everyone will recognize that a quotation I have mangled is from Shakespeare, who put it in the mouth of Hamlet: "When the wind is southerly, I know a hawk from a handsaw." Even more will know Genesis, 1: 27: "So God created man in his own image, in the image of God created he him." Both the pious and the impious have observed the probability of the inverse statement. My favorite statement in that vein is Goethe's

> *Wie einer ist, so ist sein Gott,*
> *Darum ward Gott so oft zu Spott.*

which might be rendered:

> From one's self-image God is born.
> And thus God often merits scorn.

You sometimes hear the argument that Christianity must be true because it has survived so long. The logic is absurd, of course, but it is also amusing because in that case the truest religion must be the oldest, which would put Christianity quite out of the running. I am not sure what would be the true religion by that criterion, perhaps that of some head-hunting tribe in central New Guinea.

You also sometimes hear that we *must* have dogmatic religion of some sort (in the U.S.A. the flavors offered are usually Catholic, Protestant, or Jewish) because otherwise no one would behave himself. I don't have as much faith in my fellow man as I used to, but thank god (my god, of course) I still have more than that. I know large numbers of people who are not dogmatically religious, and on the whole their behavior is better than that of the average churchgoer. Historically, it might be a toss-up, but some fanatically dogmatic characters have been very ugly customers indeed, and some atheists, agnostics, and deists have led lives of utmost purity and ethical service. By the way, many Christian patriots of today have conveniently forgotten that the deists, who included some of our country's most revered founding fathers, were specifically anti-Christian and antidogmatic in religion.

You will find more about my views on theology in chapter 11 of my book, *This View of Life* (New York: Harcourt Brace Jovanovich, 1964), and my views on ethics in chapter 10 of *Biology and Man* (New York: Harcourt Brace Jovanovich, 1969).

6
Getting Started

It is the first of all problems for a man to find out what kind of work he is to do in this universe. — Carlyle

Thus far I have not followed the chronological course usual in biography but have hopped lightly about from subject to subject and time to time. It is true that in an imaginatively literal sense material events follow a single and linear course. But a life is not composed merely of such events. It is composed also of thoughts, memories, consequences, and precedents piling ever higher with the years, and so life becomes a rich compound, far more complex than a chronicle and organized along different and multiple axes.

Now I will turn for a time, at least, to a more temporal sequence. I find that it is true that as one grows old one still remembers earlier events with considerable vividness and tends to savor them. That accumulation is one of the riches of a full and long life, but my life is still so active that I have only limited leisure for reliving the past. Much more could be said about my childhood and boyhood, and that involves some temptation, but there is too much else to say, and I shall resist that particular temptation. Instead of returning now to childhood I shall turn to a chronological description of my early professional years.

Preparation for a profession has already been sketchily outlined in the chapter on my formal education. My first professional job indeed antedated the end of those formalities, for in 1924, while still in graduate school, I was employed as a summer field assistant to Dr. William Diller Matthew of the American Museum of Natural History, New York. Matthew was at the height of his powers and his fame as one of the greatest vertebrate paleontologists, and in his company I was awed, instructed, and had some rough edges polished off. Much of the time we shared a room in a village boarding house in the Panhandle of Texas and so became as well acquainted as possible for a man of fifty-three at the top of his profession and a neophyte of twenty-two who hoped to enter it. I was trained by Matthew at least as much as by my major formal professor, Lull.

One of Matthew's oddities (he did have several) was that he had not learned to drive a car (yes, even in 1924 it was odd for a man of his age not to know how to drive). A requirement, in fact his major requirement, for a field assistant was to operate and maintain a Ford model T. In fact I had never driven a car either; college and graduate students did not have cars in those poverty-stricken and primitive days. Of course I was not going to forgo what I rightly saw as a tremendous opportunity just for the sake of a quixotic adherence to the truth, so I lied.

The car was delivered to us in a crowded garage in Amarillo. Someone else cranked it. I had seen someone drive a Model T once, so my lie was not absolute: I had a vague notion of what to push, pull, and turn. Off we went somewhat jerkily into the traffic, and I believe that Matthew never realized how close he came to ending his career at that high point. In fact, although he lived to be only fifty-nine he had several more years at the American Museum and then a good start on a new career at the University of California.

We talked endlessly, or usually Matthew talked and I asked questions. At his worst, which was still pretty good, he was given to *ex cathedra* pronouncements, but more often he was factually instructive or gave good advice to a hopeful tyro.

Regarding a scientist who had achieved fame and fortune by what seemed quite shaky accomplishment:

> W.D.M.: Such things separate the sheep from the goats. The goats turn to science for fame and material awards, whether or not on false pretences. Only a few have a genuine thirst for the truth.
> G.G.S.: I hope I am one of them.
> W.D.M.: Everyone claims good motives. That is too easy. Only time will tell.

Again on the aims of science and scientists:

> G.G.S.: I suppose for a start you try to determine whether something is so. Any truth must be valuable.
> W.D.M.: (paraphrasing Darwin, as he probably knew but I did not): No. It is not enough to seek knowledge for its own sake.

One must *know* beforehand that the knowledge is worth seeking. You could count the grains of sand on Coney Island. The result would be true and an addition to knowledge. But anyone who devotes himself to such pursuits wastes his time and deserves nothing from society.

Matthew's father, a Canadian, as Matthew remained throughout his adult lifetime in the United States, had been an amateur geologist of some note. Matthew was fond of quoting his father's maxim, "Do the next thing." Matthew advised me to get into some institution that had no paleontology and to build it up myself from scratch. I thought, but hadn't the nerve to say, that Matthew himself certainly had not followed either his father's maxim or his advice to me. In fact only three years later Matthew urged me to follow his example and not his advice, and helped make it possible for me to do so.

The principal result of our labors in the field were the discovery and collection at Mount Blanco, near Crosbyton, Texas, of remains of a graceful little three-toed horse (*Nannippus*) and two skeletons of a large one-toed horse that Matthew considered a missing link in the ancestry of modern horses and named *Plesippus*.

Matthew left before the end of the summer and I spent the rest of the field season with a man of about my age but with more field experience, Charles Falkenbach, collecting in the Rio Grande valley around Española, New Mexico. I thought that I was still collecting for the American Museum but later learned that in fact I had been working for Childs Frick. This Frick was an amateur paleontologist who had inherited a fortune from his father, Henry Clay Frick, of Homestead fame or infamy depending on your views about sociology. Childs Frick usefully expended much of that wealth on amassing a great collection of fossil mammals. Later, when I was on the staff of the American Museum, Frick took a tremendous dislike to me. I could understand someone not liking me, but I did not reciprocate the dislike and I was (and still am) honestly unaware of having given any cause for it. His dislike was so great that Frick threatened to break off financially beneficial relationships with the museum unless I was fired. To my surprise the director refused to fire me, and to my somewhat lesser surprise Frick did not carry out his threat. His magnificent collection, which I was not allowed to see, did go to the American Museum after his death.

After 1924, onward at Yale until 1926, commencement in the literal as well as scholastic sense for me, and my departure for Europe with a

Ph.D. and a National Research Council and National Educational Fellowship for a year of postdoctoral research.

In preparing my doctoral dissertation I had studied all the specimens of Mesozoic mammals, mute and fragmentary relics of the first two-thirds of mammalian history, that were available in American museums. Now my main project was to do the same in European museums. There most of the specimens and therefore my central interest were in the British Museum (Natural History) in Cromwell Road, South Kensington, an institution completely distinct from the British Museum in Bloomsbury, housing the national library and the Elgin marbles, among innumerable other things.

As my then wife preferred a villa in the south of France to a flat in London with me, most of my modest stipend went for that purpose and my own circumstances were decidedly straitened. The keeper of geology (what we call *curators* are *keepers* in England), F. S. Bather, an invertebrate paleontologist, received me cordially and helped me to settle in. Lodging was found on Trebovir Road in Earl's Court, not too far from the B.M. (N.H.), a Spartan attic room in a second-rate "residential hotel" (British euphemism for a respectable but poor boarding house). There was a communal bathroom only one floor below my garret and my room had a small gas heater kept alive by a shilling in the slot when I had one. I have made that sound rather dreary, and so it was as seen now in retrospect, but in fact I was quite happy there.

I settled down at the B.M. (N.H.) in working space in the office of Arthur Tindell Hopwood, in charge of fossil mammals, who was then "Arthur" but became "A. Tindell," which I fear was characteristic of him in a later phase. At the time Arthur and I found each other quite congenial and soon were intimate friends. He was not London-bred and so had made himself more acquainted with the great city than most born Londoners. We worked as long hours as possible — overtime work was not only discouraged; it was forbidden — and then took off to meetings or to exploration of the city. The meetings were especially those of the Zoological Society and the Linnean Society, but also included on occasion others, such as the Royal Society and the Geological Society. Eventually I was to become an honorary member of some of those societies, but in 1926–27 I did not even dream of such an apotheosis and listened to the words of my elders with due humility. I did meet many of those already rather awesome characters and a few of them became lifelong friends, notably Wilfrid Le Gros Clark, Julian Huxley, and D. M. S. Watson. Unfortunately that was not to be true of Arthur Hopwood. He gradually lost the ambition and zest of youth that we had shared, and as Arthur became Tindell we drifted apart. The last time I was in London before his death he did not have time to see me. It had then been years since he had done any of the research we had both so eagerly planned in rosier days.

In the course of our rambles during those younger days, I fell in love with London. I have been in London many times since but have retained the feeling from my first, longest stay. Here a somewhat mis-placed epigraph is again apt:

When a man is tired of London he is tired of life. — Samuel Johnson as reported by Boswell

I am not tired of London, and although my first wife refused even to visit the city my second (and definitive) wife fell in love with it at first sight.

There were a few relevant specimens scattered here and there else-where in England and on the continent. In England I also visited Oxford, Bath, and York and found them all fascinating beyond the fact that a few Mesozoic mammals were preserved in each. Moreover, some of the English countryside was seen on excursions with Sir Thomas Holland, who was kindly and generous to a young acquaintance, and on a visit to Sir Arthur Smith Woodward, the Nestor of British verte-brate paleontologists and long keeper of geology at the B.M. (N.H.). He had left London in a huff and never again entered the museum after he was passed over in the appointment of a new director there, but he was a great and, again, a kindly man to me and other tyros.

I managed a tour on the continent: Paris, Lyons, Geneva, Basel, Tübingen, Munich, Berlin, Frankfurt am Main — which sounds like quite a lot but I found distances in Europe ridiculously short, a com-mon first reaction of western Americans. This trip added bits to my Mesozoic mammal study, but was more noteworthy for my meeting paleontologists in those cities: Boule, Depéret, Revilliod, Stehlin, von Huene, Stromer, Dietrich, Edinger, all dead now and, if my current graduate students are typical, mostly forgotten. (I have matched the names to the cities.)

Before leaving London in the fall of 1927 I had completed a lengthy monograph on all then known European Mesozoic mammals, which in the curious way of the B.M. (N.H.) was published under the false guise of a catalogue of the specimens in that institution. That appeared (1928) before Yale's publication of my dissertation (1929) and so was my first monograph-length publication. I had also finished some other odds and ends such as most of a catalogue of Cenozoic fossil marsupials (that really was a catalogue in the sense of the *Fossilium Catalogus* in which it appeared.)

This was a crucial point, indeed the most crucial point in my professional development. Personal factors of considerable delicacy are involved, but some explanation must be given, and I shall attempt to do so as far as necessary, frankly but briefly and without much possibly offensive detail. In 1926 I had been offered continuing support at Yale but had declined because the available year of research and study in Europe was of more immediate interest at that point. In 1927, while I was still in England, I was given to understand that an appointment at Yale was still available and was urged to apply. I did so, but to my growing confusion and finally consternation my application fell into limbo, without acknowledgment or reply. Matthew came through London and told me that there might be an opening at the American Museum. (In fact, he was leaving that institution and had recommended my being brought in at the bottom of the department when he went out at the top.)

Lull also visited London, and he told me that he had been commissioned to investigate my family affairs. My then wife had informed friends on the Yale faculty that I was not supporting her and that she was in desperate need while I lived in luxury apart from her. Documents in my hands demonstrated to Lull that in fact she had refused to live where I was obliged to and that most of my small income went to her and for her villa, while I lived in a garret on a pittance. Upon receiving Lull's report, Yale informed me that all was forgiven and offered me a definite appointment. I declined to go where prospective associates had been so willing to believe me a scoundrel. At this desperate point, a solid offer from the American Museum arrived, and I accepted it. Thus the next thirty-two years, the greatest part of my professional life, came to be spent at the American Museum rather than at Yale.

At this point I will briefly break the chronological sequence in order to finish with this unpleasant topic. In 1942 a new director at the American Museum (he happened to come from Yale) abolished paleontology as a distinct subject, to my great distress. At about the same time Yale again offered me a position, now a good professorship, and I accepted provisionally while I was off in the army (1942-44). The provision was that I would go to Yale unless the American Museum changed its policy toward my science. It did so, and I felt ethically obliged to remain at the museum on my return from the army, now as chairman of a Department of Geology and Paleontology.

Now back to November, 1927, when I joined the Department of Vertebrate Paleontology of the American Museum of Natural History as assistant curator, a raw recruit but rapidly gaining experience and a measure of expertise of various sorts. Matthew had left, off to Berkeley and to first experiences of teaching, where, he told me, he took some time getting used to facing what seemed to be a large audience of bare feminine knees. Henry Fairfield Osborn, recently content with being president of the museum, now made himself also once more *de jure* chairman of the department. Barnum Brown was curator of fossil reptiles; Walter Granger, curator of fossil mammals; Charles C. Mook, associate curator of fossil reptiles; and William K. Gregory, research associate in paleontology. Jannette May Lucas (daughter of a former director of the museum) was assistant librarian in charge of the department library, named for Osborn and at that time distinct from the museum library. Rachel Husband (later Nichols) was an assistant, later to be in charge of the Osborn Library and of cataloguing. There was an excellent field and laboratory staff headed by Thomson, who was christened Albert but was always called Bill, and including Otto Falkenbach, George Olsen, Peter Kaisen, Carl Sorensen, and Charles Lang, all names worthy of record in the annals of paleontology.

Granger was then almost entirely involved in the long work of the Central Asiatic Expeditions, so-called although Granger was then at least equally concerned with his work in Yunnan, which is in southwestern China and not in central Asia. Except on somewhat ceremonial occasions, graced by Osborn, Brown was the *de facto* department head, and it was he who put me to work. He thought that all young squirts should start at the bottom, so he started me at preparation in the laboratory. I was twenty-five, father of a family, and had been working at a professional level for several years, but he was fifty-four and incomparably more experienced. I quite enjoyed my initiation in the laboratory. I liked and admired the men there, and it was interesting to prepare some relatively primitive mammals, which I was later allowed to study and publish.

In complete contrast to my relationship with Walter Granger, that with Brown never went beyond a courteous but distant acquaintance to become friendship. I was never in his home, never addressed him other than as Dr. Brown, and was never addressed by him except as Dr.

Simpson. Granger and I were soon Walter and George, and he became virtually a loved second father to me. Both Granger and Brown published valuable research but both preferred and were better known for collecting, Granger mostly but not exclusively of fossil mammals and Brown mostly but not exclusively of dinosaurs.

Osborn, nearing the end of a long and highly productive career, was a great man and fully conscious of that fact. His surprisingly naive and quite justified pride in his social and scientific prominence was resented by some out of their own fear of inferiority, but I found it amusing and indeed rather endearing. There are many anecdotes, and I tell just one as a fair example. One day he came into my office to present me with a copy of his massive, two-volume monograph on titanotheres. Across the flyleaf he wrote in his large and elegant hand:

To George Gaylord Simpson
with the appreciation
and best wishes of
Henry Fairfield Osborn
_____

As this was in heavy, flowing black ink, I reached for a blotter, but he stayed my hand and said, "Never blot the signature of a great man."

He was uniformly cordial and helpful to me even when, as happened, some of my work contradicted his. He required civility but not subservience from all those working under him as president, chairman, and curator within the museum, but some were not civil behind his back. He was always known as The Professor and so addressed, even though when I went to the museum it had been seventeen years since he had been an active professor. (Active professors in the museum, eventually including me, were never addressed as professor; there was never more than one Professor there.) Mrs. Osborn, Lucretia née Perry, was also properly conscious of her social position, but she favored me because she had learned (not from me) that I had seventeenth-century ancestors in Fairfield, Connecticut, whence came The Professor's middle name.

The Professor died at the age of seventy-eight in 1935. Not long thereafter one of his sons approached me to write a biography of him with the cooperation and support of the surviving family. I agreed with one proviso: that there would be no revision or deletion by anyone. The proviso was accepted on the spot (we were lunching at an exclusive club), but that was the end of the matter. There is no biography of Osborn, and that is a pity because he was the last of the titans who for a time dominated and in themselves symbolized the science of vertebrate paleontology. Preceding him were Marsh and Cope, both now with adequate biographies, that of Cope by Osborn. Nowadays the science is far more active than in Osborn's day but there are so many excellent and productive practitioners of it that no one can stand out as such a colossus.

When I went to the American Museum I had almost completed the study of all the Mesozoic mammals then known. I soon did complete, as far as then possible, study of those most recently discovered, found in the Gobi Desert by Granger and his associates on the American Museum's central Asiatic expeditions. The chances of finding more in the U.S.A. or elsewhere were judged too slight for the museum to gamble even a hundred dollars or so on the attempt. Thus I was out of a specialty and had to extend my scope, which I was glad to do and had indeed already started to do in a relatively small way. A logical extension, having studied the oldest known mammals, was to go on with the next-to-oldest, the more varied, more abundant mammals of the Paleocene, first epoch of the Cenozoic era, the so-called Age of Mammals.

At hand was a collection of Paleocene mammals made partly by Barnum Brown but mostly by Rachel Nichols from a coal mine at Bear Creek, Montana. Rachel's work was then a highly unusual example of a woman's collecting fossil vertebrates in the field, even though some of the major discoveries of the early nineteenth century had been made by a woman, Mary Anning. (Now there are a number of excellent female collectors and general field workers as well as researchers.)

These were among the specimens that Brown had set me to preparing, and having prepared them I also studied them and published the results. Another task assigned to me as soon as I went to the museum was to take over identification and other study of a large collection of much later, almost recent but prehistoric (late Pleistocene) mammals being made in Florida. This was the hobby of Walter W. Holmes, a retired businessman from Waterbury, Connecticut, who had a winter home in St. Petersburg. He happened on some curious fossil bones near that city, became fascinated by them, and went about amassing a large collection of them, which he first sent for identification and later gave to the American Museum.

I visited Holmes in Florida, determined the geological circumstances of his discoveries, and soon spread my attention farther over the state, often with Herman Gunter, then the state geologist. My studies not only increased knowledge of the late fauna exemplified by the Holmes collection, but also included much older mammals from the Miocene and Pliocene (the two epochs of the Age of Mammals just preceding the Pleistocene or Ice Age). Most interesting was my discovery of an almost complete skeleton of a sea cow, quite different from the living manatees present in Florida. The skeleton, well prepared and mounted, can still be seen at the American Museum, or at least could when I was last there a few years ago.

Such occupations and others too numerous or too minor for report here saw me through 1928 and the spring of 1929. Then I completed the process of getting well-started by taking my own expedition into the San Juan Basin of northwestern New Mexico to collect mammals and other fossils from the very oldest fauna of the Age of Mammals, the fauna that we call Puercan after one of the several Puerco ("Muddy") Rivers of New Mexico. It was not a very grand expedition: besides me it comprised a college student (who did not go on in paleontology) and a local (Farmington) young man as camp cook and helper.

That was long before the oil and gas development in the San Juan Basin. The greater part of the basin, all but its margins, was wild, unfenced territory, without roads unless you count the tracks of Indian wagons, sparsely occupied by a few Navajos and still fewer traders with the Indians (not with the tourists, as now). It was wonderful. As mentioned before, I had been in New Mexico in 1924, but that summer in the then incomparably wide-open spaces of the San Juan Basin started a love affair with northwestern New Mexico that was to last for many years.

We traveled cross-country in an old, screen-bodied Dodge truck, inherited in utter decrepitude from Barnum Brown. We camped sometimes on the ground, usually in a couple of tents, hauled water from trading posts or springs except to one camp at a spring (Barrel Spring), became blistered as we worked in the sun all day, and slept lulled by coyote songs all night. The season began and ended with minor disasters. We put up our first camp near a place (then a trading post) usually indicated as Kimbetoh although kinne-bi-toh is somewhat nearer the Navajo, which means "sparrow hawk, his water" (that is, spring, although in fact water was obtained by digging in the bed of the arroyo). We were just relaxing after a long, hard day when a whirlwind, a large dust devil in the local Anglo patois, came through camp, ripping up our tents, and spreading everything we had over a half mile or so of sage brush.

The end of the season was at least equally dramatic, although I was not there. I had gone in to Albuquerque on business as we were nearing the end of summer and of the late summer rains characteristic of the Southwest. Floods delayed my return, but the two others packed up camp and started out. In an arroyo the truck became stuck, which was not unusual, but it was unusual, indeed unique, that the drive shaft broke. The young men wisely unloaded everything in a hurry and carried it all out to the high rim of the arroyo. They had no sooner done so than a tremendous flash flood came down the arroyo. As they say nowadays, the truck was totaled. But we had a good collection, and we got it out and back to New York.

Notes

§ During my teens and twenties I wrote a great deal of doggerel, often humorous in intention if not in fact. Later I did not so much become more sober as more inclined to express my sense of the ridiculous in other ways. I had not emerged from the earlier phase when I was first in England, and mention of my visit to Bath suggests to me a sample by way of indication that my interests and reactions were not wholly wrapped up in peering at diminutive teeth through a microscope. I was entranced by Bath and its legends, which extend back from the days of Beau Brummel into those of the Romans and still earlier. Among the earliest is the legend of Prince Bladud and the discovery of the curative powers of the spa at Bath.

> Prince Bladud, he
> Had leprosy
> About nine hundred years B.C.
> His father great
> Was heard to state
> He'd have to give the prince the gate.
>
> His money spent,
> The poor prince went
> From bad to worse, and not content
> To starve and die
> He thought he'd try
> The post of keeper of the sty.
>
> Prince Bladud, he
> Gave leprosy
> (Misplacéd generosity)
> To all his swine —

Had they been mine
I wouldn't have thought Blad so fine.

They found one day
Along their way
A place where steaming waters play,
And where they flow
The ground is low,
Black mud abounds and rushes grow.

O lovely mire!
O heart's desire!
It set the suilline souls afire!
They wallowed through,
The prince did too —
Those waters fixed them good as new!

Prince Bladud, he
And leprosy
At last had parted company.
The blessèd spring
Whose praise I sing
Had made him fit to be a king.

That is a sufficient, more than sufficient, sample of the sort of thing
that amused me when I was twenty-four.

§   In reference to the biography of Osborn, or rather the lack of one,
     it should be noted that he did publish a volume in the guise of an
autobiography. My copy was autographed with an unblotted signature
and with the notation that the date, "August 8th 1930," was his sev-
enty-third birthday. Nineteen pages of the volume are more or less
autobiographical but all the rest are occupied by his bibliography up to
1929 in two different forms, one chronological and one topical, and by
lists and accounts of all of the many honors bestowed upon him. His
bibliography at that time, by the way, had 801 entries.

§   Mary Anning, whom I had occasion to mention in this chapter, is
     an odd character well known in the history of paleontology. Her
father, Richard Anning, was a carpenter in Lyme Regis on the southern
coast of England. As a sideline he collected and sold fossils as curios. As
a child his daughter Mary was an interested companion on his fossil-
hunting excursions, and some time after he died in 1810, when she was
only eleven years old, she undertook to carry on the fossil business on

her own. While hardly more than a child she found the most complete ichthyosaur to have then come to scientific attention. She employed men to dig it out and sold it to the Reverend William Conybeare, who published a scientific notice on it. Later, at the ripe age of twenty-two, she discovered a similarly complete plesiosaur skeleton, also sold to Conybeare and published by him, and when twenty-nine she found the first recognized British pterodactyl. Her usual stock-in-trade was fossil shells and other curios sold to nonscientific collectors (Lyme Regis was and is a resort town). She did find other ichthyosaurs and plesiosaurs, however, and sold one of each to Dr. (medical) T. B. Watson, who presented them to the Academy of Natural Sciences in Philadelphia. Mary never married and we can charitably ascribe this to her greater passion for fossils in spite of the fact that a contemporary described her as "a prim, pedantic, vinegar-looking, thin female, shrewd, and rather satirical in her conversation." (Women paleontologists today are mostly shrewd, sometimes satirical, but merit none of the other adjectives.) Mary Anning contracted cancer and died at the age of 48.

While footnoting equality for women, I may also mention that it was perhaps a woman who found the first dinosaur. The discovery is often credited to a British physician, Gideon Mantell, in 1822, but according to an early account it was his wife who saw it first and called his attention to it. (My wife, Anne, although a psychologist has also made important finds of fossils while we were in the field together.)

References

Camp, L. S. de, and C. C. de Camp. 1968. *The day of the dinosaur.* Garden City: Doubleday. This is a good example of popularization of science — not all dinosaurs; it includes comments on Mary Anning.

Osborn, H. F. 1930. *Fifty-two years of research, observation and publication, 1877–1929: a life adventure in breadth and depth.* New York: Scribner. This is the pseudoautobiography referred to in my second note.

Simpson, G. G. 1928. *A catalogue of the Mesozoic mammalia in the geological department of the British Museum.* London: British Museum (Natural History).

———. 1942. Memorial to Walter Granger. *Proc. Geol. Soc. Amer. for 1941:* 159–72.

———. 1944. Osborn, Henry Fairfield. *Dictionary of American biography,* vol. 21, supplement 1, pp. 584–87. New York: Scribner.

7

How We Knew Where to Dig

I never met a man I didn't like. — Attributed to Will Rogers

I am convinced that the only people worthy of consideration in this world are the unusual ones. — Attributed to a scarecrow by L. Frank Baum

When I was considering how to get into this chapter, going on chronologically into 1930 and beyond, those two epigraphs came into my mind in conjunction. If Will Rogers really meant that, his remark is what he called politics in the musty slang of his day: applesauce. Anyone who has met many people and liked them all is at worst a fool and is at best lacking in taste and judgment. The scarecrow, on the other hand, was a snob because his ideas of the unusual were too bizarre to apply in a real world. Yet putting the two remarks together made me realize that I have never met anyone who was not unusual by mundane standards, whether likable or not. And that thought carried me back to Argentina and especially to its southern extension, Patagonia, which I first entered in 1930.

Further experience has convinced me that no two people, not even identical twins, are quite alike. Thus everyone is unusual in some respect, although it is true that some are more unusual than others — or, to reapply the old wisecrack, some people are more equal than others. The relatively few inhabitants of wind-swept Patagonia in the 1930s were surely more individual than most, more decidedly one of a kind.

Reaching back into my Patagonia memories I recall examples so numerous that it is difficult to choose, but I have just come across an account that I wrote there about my friend Don Mariano, born in

Algiers of partly French and partly unknown parentage but a longtime resident of Patagonia. He will do nicely as an example, certainly not a typical example because the point is that no one is typical, especially not in Patagonia.

An Indian, a half-breed, another North American, and I were squatting on the floor around a small fire in the *tapera de López* taking *mate* when Don Mariano came into my life. The tapera de López was an abandoned hut in the dead center of Patagonia. *Mate* (pronounced more or less mah'-tay) is the universal tealike beverage of Argentina, sometimes called "Paraguay Tea" in English.

"*Buenas!*" called a voice from outside.

"*Buenas! Pase!*"

Don Mariano entered. About five feet tall, thin to the point of emaciation, clothed in a way noteworthy in Patagonia only in being more exiguous, ancient, and tattered than usual. His face was in the condition, almost universal in Patagonia, of one who does not definitely have a beard and yet seems never to shave (or to wash, for that matter). He lived alone in a deserted landscape and had bottled up a lot of conversation that came flooding out in idiomatic but oddly accented Spanish.

"So you are Yanquis, eh? Hunting for bones, eh? Well, there should be a lot of them around. Sure to be. Richest place on earth, central Patagonia. Everything occurs here. Bones too. *Pucha!* [a vulgar euphemism for an even more vulgar expression] Millions of them. No doubt about it. You're in the right place. But why haven't you been to see my farm? What! Didn't know there was a farm in Central Chubut? Why, it's the showplace of Patagonia.

"What do I grow? Everything! Wonderful soil. A little water, *mucho trabajo,* and the desert blooms. Fruits and nuts, mostly. There's the real future of Patagonia. I have to laugh when I see my neighbors raising sheep. [There was a neighbor only ten or fifteen miles distant.] Wasting time on those animals when the soil is a regular gold mine.

"They think I'm crazy. They're the crazies. Raising sheep! And what happens? What happens, do you ask? A dry year or a cold year or a windy year happens. And then what, do you ask? Then the sheep die. Mucho trabajo, no wool, no money. Or perhaps a good year. And then? The wool is so plentiful it's not worth anything. Still mucho trabajo, no money. Not even for *vicios.*" To a Patagonian a *vicio* is not a vice but anything not absolutely necessary to sustain life, even the *yerba* to brew *mate,* and tobacco, perhaps some sugar, and a little wine if the year is especially good.

"What have I planted? Well, for example three hundred almond trees, one hundred apple trees and lots of olive trees. You can make oil, stock feed, anything. On almonds and apples a man can live like a king. Well, you know, not all of them lived. True, most of them died. No water, and the *cuys* ate the young trees." Cuys are Patagonian wild guinea pigs.

So on for an hour or so, but he would not stay for dinner, assuring us that his farm provided him with all a man could eat and more. So much better than meat, meat, meat all the time. (Most Patagonians eat practically nothing but mutton year in and year out, as we did, too, while there.) So we met the only vegetarian in Patagonia. However, a couple of weeks later, on his third or fourth visit, he did stay for dinner and wolfed down a whole quarter of roast mutton.

"Meat is good from time to time, but foolish people don't realize that fruit and nuts are better for a steady diet."

Between bites he told us something of his travels, Algiers, Marseilles, and even the United States, which he had visited as a sailor.

"Nueva York. I remember it like yesterday. That sort of statue in the harbor there. What do you call it? The big lady in a nightgown holding up a bundle of firewood. Yes, *Libertad!* That's it. Only I didn't see her. When we arrived I was in the brig. When we left it was night, and I was drunk anyway."

In spite of a somewhat peripatetic past, he had been in this spot so long that he had no idea of anything that had happened more than twenty or thirty miles from there since about 1900.

Seeing a can that had contained a popular brand of tobacco, he face-tiously remarked, "So the King of Belgium is traveling in Patagonia?"

"That isn't the King of Belgium. It's the Principe Alberto, husband of the Reina Victoria of England."

"So England has a queen now! That *is* news!"

"No, the Reina Victoria is dead."

"Well, I thought so. We're not so out of the world as that. Even here we know that Eduardo Sétimo is king of England."

"Eduardo Sétimo is dead, too. His son Jorge Quinto is king now."

"So Eduardo is dead. Poor man! Well, kings have to die, too. You *do* bring news."

"It happened some time ago."

"Well, you know how things are here. Newspapers don't arrive every week, even if we could read them."

I could quote Don Mariano with pleasure for many more pages, but I won't. I will just add that I did later visit his "farm." There were three cabbages, which had failed to head. There was a scrawny tree with one, sole apple. There were perhaps a dozen scraggly slips that Don Mariano said were olive trees. Half of them were leafless and dead, the rest plainly moribund. None had ever borne olives. And that was all.

"You know," Don Mariano said hesitantly as I left, "everyone around here thinks I am crazy. What do you think?"

I said only, *"Que le vaya bien, Don Mariano."*

That was in fact during my second expedition to Patagonia, in 1933–34, and not my first, in 1930–31. That first Patagonian expedition was decidedly one of the most important events in my life, so it should be emphasized here. But I have already written a whole book about it: *Attending Marvels* (for further bibliographic details see the references at the end of this chapter). It would be pointless to try to sum up the book in a few pages here, and I am not inclined to do so. I will simply outline the facts and let it go at that.

In the northern fall, southern spring, of 1930, Coleman Williams and I went by ship to Argentina on the First Scarritt Expedition. Coley was a young man who had been instrumental in putting us in touch with a wealthy New Yorker, Horace Scarritt, who provided the funds for the expedition. In those remote times there was no government support at any level for scientific research outside an institution directly incorporated in a branch of the government, nor did the American Museum (a nonprofit but private corporation) have any budgetary funds for work outside its own building. For a project like this we were dependent on philanthropic angels and for a few years Horace Scarritt was my much appreciated angel.

From Buenos Aires we went on a small oil tanker to Comodoro Rivadavia, the metropolis of Chubut Province in Central Patagonia. From there we traveled inland by truck, over wool-wagon tracks or no tracks at all, hunting, finding, and collecting fossils. Some local people worked with us. By far the most valuable was Justino Hernández, a lad who had never been out of central Patagonia, cheerful, hard-working, highly congenial, son of an Araucanian Indian father and a Lithuanian mother. (He spoke only a Patagonian dialect of Spanish, which they call "Castellano.")

In May, beginning of the southern winter, we drove with great difficulty from Comodoro to Buenos Aires, where we shipped our fossils off to New York. Coley also went to New York to work over the fossils in the laboratory at the American Museum. I spent the winter in Buenos Aires and the nearby city of La Plata, studying the extensive fossil collections in the museums of those two places. The idea, eventually carried out, was that I would combine knowledge of those collections with my own excellent collection to write a monograph on the earliest South American faunas of the Age of Mammals, the Cenozoic in more technical geological terminology.

Fossil mammals had been discovered by Charles Darwin in 1833 at several places in Argentina. Most of those fossils were described by Richard Owen, then a somewhat older and paleontologically more sophisticated friend of Darwin, later an implacable opponent because of Darwin's espousal of evolution based in part on those very fossils. Darwin's Argentinian fossils were not very old, geologically speaking, mostly from toward the end of the last Ice Age (the Pleistocene) and datable in thousands rather than millions of years. Later, in the latter part of the nineteenth century, it was found that the rocks of Argentina, and most particularly those of Patagonia, contain fossils covering almost the whole extent of the Age of Mammals, from about 60 million years ago on.

That discovery was due largely to two extraordinary brothers, sons of humble Italian immigrants, Florentino and Carlos Ameghino. Most of the fossils were collected and the stratigraphic observations on the succession of the various faunas made by Carlos. Florentino studied, named, and described the fossils, often in French with the aid of his French wife, Léontine née Poirier. Carlos, whose most vigorous years were spent in what were then wilds, never married. With some exceptions, their large collections eventually came to rest in the changeably named national museum in Buenos Aires. (Its current full name is Museo Argentino de Historia Natural "Bernardino Rivadavia.") Most of my work in the southern winter of 1931 was devoted to studying the older fossil mammals of the Ameghino Collection in that museum.

The museum was then crowded into a small building on the Calle Peru in the center of the city. (It soon moved to a large new edifice in a more outlying area.) Florentino Ameghino had died in 1911 in his fifty-seventh year, while I was a child and quite unaware of the Ameghinos, but in 1931 Don Carlos was still living although he was then aged and in poor health (he died about a year later). In remissions of the illness that proved terminal, he would come to the museum and drink *mate* with me in gaucho style, that is, through a shared silver tube (*bombilla*) from a gourd (the "*mate*" for which the drink is named) and not in tea-cups in the degenerate city fashion. Although he was literate, most of his field notes were in his mind rather than on paper, and to my delight most of them were still there and lucid. Talking to him about his years in Patagonia was like what I think it might have been to talk to Darwin about his (fewer) years on the *Beagle*.

The active student of fossil mammals in Buenos Aires was then Lucas Kraglievich, friendly, cooperative, and able. Unfortunately he soon thereafter let himself be involved in a plot against the administration of the museum, went briefly into virtual exile in Uruguay when the plot failed, and died in 1932 at only forty-five. In 1931 I shared many a drink of *mate* and much edifying conversation with him, too.

The principal collection relevant to my aims in La Plata's then more elegant museum, now still standing with little modification, had been made by Santiago Roth, Swiss-born but long resident in Argentina. He had died in 1924 and I never met him, but in 1931 his apparently somewhat prickly character was well remembered. During my stay in La Plata, shorter than in Buenos Aires as many fewer specimens there were relevant to my project, the fossil mammals were under the charge of Angel Cabrera. Cabrera was born in Spain and had gained an excellent reputation there as a student of living mammals before he emigrated to Argentina and extended his studies to include fossil mammals. His flawless Castilian fell strangely but dulcetly on my ears, accustomed to the speech that the Argentinians spell Castellano but pronounce ca-stay-zhah-no. It is in fact a dialect quite distinct from the Spanish of Castile. In addition to being an erudite zoologist, Cabrera was a competent artist who illustrated his own publications always accurately and sometimes charmingly.

With the change of hemispheres, I arrived back in New York for my second winter of 1931. Other more routine duties had priority, but such research time as I had was spent largely on preliminary notes about some of our Patagonian discoveries. I should explain here that a great deal of laboratory work goes into making fossil vertebrates ready for study. Protective bandages put on in the field have to be removed; rocky matrix or concretion must be chiseled, scraped, or dissolved off; broken fragments must be pieced together; preservatives must be applied; special techniques such as grinding serial sections or making internal casts of skulls may be called for. Thus with a large collection like ours from Patagonia it may be years before all the specimens are ready for study and a definitive publication based on them. In the meantime it hastens the dissemination of knowledge if preliminary notes are published on particularly noteworthy specimens as laboratory work on them is completed.

In the meantime another large project came my way—I have never been able to work uninterruptedly on a single subject or thing. This project also involved early mammals of the Age of Mammals, but back in North, not South, America. Ever since 1908 the United States National Museum had been amassing a collection of mammals from the Paleocene (first epoch of the Cenozoic or Age of Mammals) of Montana. Almost all the field work had been done by Albert C. Silberling, after the first season under the general direction of J.W. Gidley, who was in charge of fossil mammals at the National Museum. Gidley had prepared most of the specimens himself, had published four preliminary notes on a few of them, and had planned a complete monograph on the collection before he died on 26 September 1931. In 1932 the then director of the National Museum, Alexander Wetmore, asked me to take over preparation of the monograph, and a cooperative arrangement was made with the American Museum.

I spent the northern summer of 1932 in the field with Al Silberling, making a map of the fossil field, which lies just east of the Crazy Mountains in central Montana, and somewhat augmenting the collection. Al, who spent his whole life in that region, except for a few excursions after fossils in other western areas, was an ideal field worker and companion. He was crazy about fossils, tireless and ingenious in their pursuit, good-natured and good-humored whatever the circumstances. Field work was not practical in the bitter Montana winters and at this stage in his life he was spending winters as an engine wiper for the Milwaukee Railroad at Harlowton, Montana. Al's wife, a Seventh Day Adventist, was a violent antievolutionist. Al, who considered evolution an obvious fact, was not at all perturbed, but the home atmosphere was not always congenial for evolutionist visitors. We were usually away in camp. (In 1935 I spent another happy and productive summer collecting Paleocene mammals with Al, this time for the American Museum.)

What with the continuing Patagonian work, other research, and various museum duties and activities, the Crazy Mountain field monograph was not completed until March 1936, and not published until August 1937. It should be noted that publication of an otherwise completed monograph or book within a year is still the exception and not the rule.

Now back to Patagonia and finally to the title of this chapter.

"How do you know where to dig?" is one of the questions most often put to bone-diggers, vertebrate paleontologists who collect as well as study fossils. Negative answers are easier than positive ones. Some kinds of rocks, such as granite, never contain fossils. Some others, such as lava or mica schist, contain fossils so rarely and, if any, so poorly preserved that it is hardly worth looking in them. So you usually look in rocks (to a geologist even sand is a rock) that were deposited as sediment on the surface of the earth or in water: shale, silt- stone, sandstone, coal, and others. But even in them if you just dug in at random you could spend a lifetime without finding a fossil bone. And you don't go around with some sort of doodlebug or mine detec- tor and dig where it indicates that there is a bone, because it practically never indicates any such thing. There are exceptions, but generally speaking a bone-digger does not start to dig until he sees a bone. So it usually boils down not to knowing where to dig right off but to know- ing where to look.

The best places to look are where someone has already found some fossil bones. Of course someone has to be first, and that can happen in perhaps a hundred different ways, sometimes considered luck or chance. But it is not luck or chance if you were systematically prospect- ing rocks in which fossil bones might occur, and it still is not luck if someone picks up one as a curiosity and eventually it is identified and traced to its origin. In 1930 – 31 we were following the first system, for the most part: searching areas where Carlos Ameghino and a few others had already found fossils.

When we went back to Patagonia in the southern summer of 1933 – 34 we found out where to dig in one of the less usual ways, by follow- ing clues in a detectival way. The first clue appeared in 1931 when we had completed our summer's work and were about to leave Patagonia. Hearing that I was interested in such useless odds and ends, a sheep- herder showed me a chunk of rock that a passerby, a wanderer or *pasiandero,* had given him. He knew only that it came from somewhere out there in the *centro,* where hardly anyone ever went. It was in fact a part of a jawbone, petrified in a way new to me and with curious teeth.

In 1933 as we worked southward from Buenos Aires we came to Trelew, one of Patagonia's few towns, originally a settlement of Welsh farmers who had fled from what they considered English persecution. Coley Williams was again with me and Justino Hernández joined us in Trelew. The area was then becoming somewhat Argentinized and was almost completely so when I revisited it many years later. In 1933 we began to find some interesting fossils there, mammals but also large numbers of ancient penguins, which were to have a peculiar influence on my life. The point here, however, is that a friend in Trelew showed me another fragment of fossilized jaw just like the one I had seen more than two years before. He didn't know where it came from, but remembered that it had been given to him by a friend from upriver (the Chubut River, one of the few flowing streams in Patagonia). Name of Espinel. A *bolichero* (proprietor of a primitive inn and country store).

So when we had finished our work around Trelew, we went upriver, a difficult but highly scenic trip, and two days later came to Espinel's *boliche*. He wasn't there, but we moved in and were happily taking *mate* when a small boy arrived, a junior Espinel. He caught a horse, rode off to round up the rest of the family, and in a couple of hours returned with *papito, mamita,* seven or eight more small Espinels, and two neighbors. (A neighbor there could be from fifty miles away.)

Yes, Papito Espinel had given a curio to my friend in Trelew. Where did it come from? No idea. A pasiandero gave it to him. Finally after cudgeling his substitute for brains:

"Oh yes, the pasiandero said something about a Turk at a place called Canquel."

"Where is Canquel?"

"Out yonder in the centro some place. Not far. Maybe thirty leagues," — about ninety miles, several days travel for our truck in a rough region without any roads.

At that time the centro of Chubut was either left completely blank on maps or was filled in with utterly fictitious detail. In that sense it was unknown territory, even though it had a few widely scattered, illiterate inhabitants and wanderers. Ours was certainly the first automobile (a Ford truck) to cross the centro and probably the first wheeled vehicle — we made our way through without previous tracks passable for wagons. We found Canquel and were not unduly surprised to learn that it is a long, lava-capped sierra located just where the only map that showed anything there indicated a broad valley.

On the east side of the sierra we also found a "Turk," almost surely not really of that origin but probably one of the Syrians or Armenians who have somehow wandered even into such remote spots. We took *mate* with the Turco while he smoked a narghileh ingeniously improvised from tin cans.

"*Güesos! Sí, hay muchos, pa' joder!*" — "Bones! Yes, there are obscenely many!" (I have bowdlerized the translation of an expression constantly on the lips of my Patagonian friends but considered highly indecent by less earthy folk.)

"Where?"

"Everywhere! Nice to find a skeleton, eh? Lots of money? In this life money is everything: everything boils down to money."

The company enlarged to include a sheepherder, a hunter of baby guanacos, *chulengos,* for fur, and a diminutive wanderer.

> *Turco:* They ought to prohibit hunting chulengos. There won't be any left.
> *Sheepherder:* A good thing, too. Then the pasture will all be for sheep.
> *Turco:* And what will poor hunters do then? With no chulengos they will have to rob. They will eat your sheep.
> *Hunter:* I don't kill sheep. (He is ignored.)
> *Sheepherder:* If they rob us, we will shoot them. Good for nothing, dirty half-breeds.
> *Justino:* Easy and gently, friend. My father is Araucanian (an Indian) and in a minute I'm going to put myself in the conversation.

The sheepherder made a half-hearted apology and no blood was shed. Conversation continued well into the night. The Turco had a magnificent vocabulary, much of it too indecent for record, but I especially liked his term for politicians: *panzistos* — "belly-ists." In the course of the talk the wanderer gave us our next, almost our last clue. He said that there was a place with lots of broken bones on the other, the west, side of the sierra.

The sierra was impassable so we had to backtrack a bit and then swing around the north end and down the west side. By evening we came to the Cañadón de las Víboras, "Viper Canyon," a narrow arroyo full of blocks of lava but with a trickle of water and three houses — dense population for that region. There was even a woman, a very young but extremely pregnant Indian girl. We worked in that area for a few days and found some good fossils but not those we were especially seeking. A relatively old Indian man, the pregnant girl's father, told us that there were many bones in an even more remote place to the south, a *rinconada* or large embayment called "de los López" for a family that had once lived there. On the day before we left the old man was killed by a fall off his horse and almost simultaneously his grandson was born. Everyone was struck by the fact that the population had not been changed by the day's events.

So with great difficulty, building road as we went, we moved into the rinconada and set camp in an abandoned hut, the tapera de López, near a spring.

We prospected with little success for a time, and finally, as I wrote in my journal for 5 December 1933:

> Going slowly across the bottomlands, I prospected and found bits of bone and one-half an armadillo scute, hopes sinking lower and lower. Near the end, as I came to the high barranca, I had just picked up a much rolled bone of larger size when up came Justino at a trot. He had finished his (camp) work and was heading for the same spot that I was. His instinct, or what have you, and my habit of overlooking no chances got us there at exactly the same moment: a perfect wealth of bones. Such a sight as I never before saw in Patagonia. Bits of fossil bones everywhere over several acres so that you cannot walk without stepping on a bone fragment at every step. Skeletons dribbling down slopes where they weathered out. We were almost delirious with joy.

Those were indeed the bones we had been trying to trail down. In place they were complete, articulated skeletons, just as the animals had died and forthwith been buried, dozens of them, at the least — a most extraordinary occurrence.

Now we knew where to dig, and we did so with a will.

Slowly I figured out what had happened. Erosion had cut through what had been a volcanic crater some 35 million years ago. There had been a small lake in the crater, and herds of strange herbivores had come in to drink, had been killed by volcanic gases, and been buried in volcanic ash falling into the lake. Besides the large ungulates, later named *Scarrittia* in honor of our angel, there were several rodent skeletons, among the oldest and most complete fossil rodents known from South America, and many small frog skeletons.

There is a strange postscript to this account of successful fossil sleuthing. On 28 June 1951 I visited the Hancock Museum in Newcastle on Tyne. There in a case, without label or any indication of its nature or source, was a fragment of *Scarrittia* that certainly came from that fossil deposit in the Rinconada de los López. When and how it got from there to this provincial museum in northern England is a mystery with no clues. It remains unsolved.

Notes

§ There is some doubt whether Florentino Ameghino was born in Italy just before his parents emigrated or in Argentina just after they arrived there. I cannot see what possible difference this could make, but Florentino's enemies, who were fairly numerous, insisted that he was born in Italy and considered this derogatory. I am rather shocked to find a standard American reference work listing him as born in Italy and named Fiorino, the Italian equivalent of Florentino. He was never called Fiorino unless, just possibly, in the first few days of his life. I am not sure when Carlos was born, but he was younger than Florentino and was certainly born in Argentina. A reference work says of him "flourished circa 1860." As he was most active in the 1880s and 1890s and lived until 1932, I doubt whether he was flourishing around 1860, not to mention the fact that his older brother was definitely born in 1854 so that Carlos cannot have been more than five years old in 1860.

There is an evaluation of the Ameghinos' work in my monograph cited at the end of the next chapter.

§ During my second season in the Crazy Mountain Field with Al Silberling, in 1935, Walter Granger and Bill Thompson (both previously mentioned) were working guests in our camp for some time. I saw a great deal of them in New York, outside of as well as during working hours, but that was the only time I was in the field with either of them. In their late years the two old campaigners used to go off together almost every summer and roam around, often in the South Dakota badlands.

§ In Mitford Mathews's *Dictionary of Americanisms on Historical Principles,* a doodlebug is defined as "Any one of various unscientific devices with which it is claimed minerals and oil deposits can be located." It is also known as a witch-hazel, a divining rod, a hoodoo stick, and by various other names. My first mother-in-law was a highly successful water-witcher. She was a nice woman, so I did not point out to her that in her part of eastern Kansas you could hardly indicate any spot where water would not be found with a little digging.

§ I contributed an account of this example of how we knew where to dig as a chapter in the Explorers Club book, *Through Hell and High Water,* cited below. It bears repeating and is here put in different words for the most part. I was active in the Explorers Club for some years and edited the Explorers Journal for a while, but I withdrew after some of my closer friends there had died or withdrawn and some of the new members did not seem to me to be explorers in my sense of the word.

§ My relief that no blood was shed when the sheepherder slandered half-breeds in the presence of Justino was not merely a way of speaking. In Patagonia every man, including visiting paleontologists, carried a large knife, a *facón,* stuck in the back of his sash *(faja).* This was used for butchering, for eating, and if occasion arose, for fighting. One afternoon while I was alone at the tapera de López an ill-willed dropper-in and I spent a long time sharpening our knives and making aggressive remarks, trying to see who would cave in first. Fortunately for paleontology, or at least for my further part in it, he did. Our last evening in the field in Patagonia in 1934 was marred by a disagreement resulting in one corpse and one walking casualty wounded in the buttocks, but that was a shoot-out, not a knife fight.

References

Chaffee, R. G. 1952. The Deseadan vertebrate fauna of the Scarritt Pocket, Patagonia. *Bulletin of the American Museum of Natural History* 98: pp. 503–62. This is a dissertation based on our collection from the Rinconada de los López. It includes a life restoration of *Scarrittia* by John C. Germann.

Cramer, S. S., ed. 1941. *Through hell and high water.* New York: Robert M. McBride and Co. This is the Explorers Club book in which chapter 21, P. 205–09, is my account of "how we knew where to dig."

Simpson, G. G. 1934. *Attending marvels: a Patagonian journal.* New York: Macmillan. An account of the 1930-31 expedition. An edition with a new introduction by Laurence Gould and some changes in the illustrations was issued in the Time Reading Program in 1965.

————. 1937. *The Fort Union of the Crazy Mountain Field, Montana, and its mammalian faunas.* U.S. National Museum Bulletin 169.

8

The Importance of South America

We see that whole series of animals, which have been created with peculiar kinds of organization, are confined to certain areas; and we can hardly suppose these structures are only adaptations to peculiarities of climate or country. — Charles Darwin

When on board H.M.S. 'Beagle' as naturalist, I was much struck with certain facts in the distribution of the inhabitants of South America, and in the geological relations of the present to the past inhabitants of that continent. These facts seemed to me to throw some light on the origin of species — that mystery of mysteries, as it has been called by one of our greatest philosophers. —Charles Darwin

The words of the first quotation above appeared in Darwin's "Journal of Researches into the Geology and Natural History of the Various Countries Visited by H.M.S. Beagle, under the Command of Captain Fitzroy, R.N. from 1832 to 1836," published in 1839 as part of the official report on the voyage and later separately as the highly popular book usually called *The Voyage of the Beagle*.

The second quotation comprises the first two sentences of the first edition of *On the Origin of Species by Means of Natural Selection or the Preservation of Favoured Races in the Struggle for Life*, the book more commonly known simply as *The Origin of Species*, published in 1859 and thereafter several times revised.

On the *Beagle*, Darwin was still speaking of animals as "created," but he was clearly finding that there was more in their distribution than meets the eye and even hinting that his observations might not be explicable by a simple concept of creation. In the opening words of his

mature masterpiece he indicated that it was the distribution of fossil and
recent South American mammals that started him well on the way to
the concept of evolution and toward a solution of the "mystery of
mysteries."

There were two points in particular that started Darwin wondering
whether the hypothesis of special creation could account for the
observed facts about faunal distribution. A view common before Darwin was that each species had been created in accordance with the climatic and other environmental conditions of the region in which God
intended it to live. Darwin noted in South America that this cannot be
true. There are environments elsewhere at similar latitudes that closely
match those of South America, for example in Africa or Australia.
Nevertheless the faunas of those other regions are almost completely
unlike that of South America.

Darwin's other pregnant observation was that, with some explicable
exceptions, even the few and relatively late South American fossils
known to him included some specimens clearly related to the living animals of that continent and no other, as well as some wholly extinct but
likewise peculiar to South America. Even in the 1830s, when Darwin
was on the *Beagle*, rational geologists had discarded the untenable view
that such fossils represented animals killed and buried by the Noachian
deluge. The first edition of Lyell's great *Principles of Geology*, which
clearly demonstrated the fallacy of Noachian geology, was published in
three volumes in 1830-33 and was studied by Darwin on the *Beagle*. (It
is a shocking example of the inadequacy of human intelligence that even
now, almost a century and a half later, there are some people who seem
really to believe that fossils are explained by the Noachian myth.) It
became increasingly clear that evolution was the only reasonable explanation of both the fossil and the recent faunas of South America.

In fact the history of the South American fauna has turned out to be
even more instructive than Darwin could foresee. This became increasingly evident as the Ameghinos carried out their long and arduous joint
task. Unfortunately, however, they missed what is now evidently one
of the most important points. Carlos made no essential errors in deter-
mining the relative or sequential ages of the fossils that he personally
collected. The correct sequence was established, for its earlier part, at
one of the most important single geological exposures in the world: the
great barranca (cliff or steep slope) south of Lake Colhué-Huapí in central Patagonia. There in plain sequence are four distinct faunas, one
above and hence later than another, covering some tens of millions of

years. With the exception of two groups (rodents and primates) that appear only in the highest, hence youngest, fauna at this locality, each fauna contains the ancestors of the next and the sequence clearly exemplifies systematic evolutionary change.

That much was recognized and, for their time, well demonstrated by the Ameghinos. (It is less important that they did not realize that the rodents and primates must have come from somewhere else.) Nevertheless, Florentino made two serious mistakes when he compared his South American fossils with those from other continents. The first was that although he had most of the sequence right, following Carlos's observations, he considered almost all the faunas, particularly the earlier ones, to be much older than in fact they are. His other basic error, a complex one, was that he considered all resemblances between groups of mammals to indicate genetic affinity. There he had bad luck, because unknown to him he was working on faunas that elaborately illustrate the phenomenon of convergent evolution. Even though two groups are of quite different ancestry, if they evolve independently in adaptation to similar ecological and behavioral conditions they frequently develop anatomical resemblances that are not indicative of genetic relationship. This happened on a particularly large scale in South America.

The combined outcome of those two fundamental errors was that Florentino Ameghino concluded that all the fossils he described were older than somewhat similar ones from other continents and were in fact ancestral to the latter. Thus the patriotic Argentinian (perhaps born in Italy) insisted that practically all known groups of mammals, decidedly including mankind, had originated in Argentina and thence spread to the rest of the world. He was hardened in this opinion and fought for it even more strenuously because some critics questioned his *sequence* of rocks and faunas, which was in fact essentially correct. A distressing example was John Bell Hatcher, an American who made a collection in Patagonia during Florentino's lifetime, and who should have known better.

It was a great French paleontologist, Albert Gaudry (1827-1908), who provided the most essential clue toward clearing up that confusion. The story starts with André Tournouër, son of a French painter who was an amateur paleontologist. André went to Argentina and was farming near the city of Mendoza when Gaudry made a suggestion to him. Gaudry's own statement merits translation and inclusion here:

One day when he [André Tournouër] was returning from Mendoza I talked to him about the discoveries of M. Ameghino and M.

Moreno in Patagonia. I asked him to undertake collecting in order to follow his father in doing honor to French science. When his father was named, he turned to me with a deep and affectionate expression. "I will try to follow him," he said to me. "I am going to go to Patagonia. The Paris museum shall have fossils." He has thoroughly lived up to his word.

André made five expeditions to Patagonia, starting in 1898. As the Ameghinos were sometimes criticized for not cooperating with other scientists — and they sometimes had good reasons not to — it is pleasant to recall that they gave Tournouër all possible help, even when it became clear that his collections were being used in Paris to refute some of their ideas. Young André made good collections at localities found by Carlos Ameghino. Between 1904 and 1909 Gaudry published five studies based on the Tournouër Collection. He advanced the view, now accepted with no serious reservation, that the known mammals from the earlier part of the Age of Mammals in South America were evolving in complete isolation from those of any other continent and along their own independent lines. The somewhat superficial resemblances on which Florentino Ameghino had relied as evidence of ancestral relationships to the mammals of other continents were due solely to adaptive convergence.

The only South American mammal that Gaudry described at any length was the strange *Pyrotherium*, convergent toward some mastodonts (relatives of elephants) on the northern continents. Gaudry planned at the age of eighty-one to go on with the study of other Tournouër specimens "if God lends me life," but God did not. He was already dead when his *Pyrotherium* monograph was published, and much of the Tournouër collection has not been studied even now, although in 1964 I published descriptions (in Spanish) of the earliest mammals in the collection. But Gaudry had found one of the keys to understanding the South American fauna: South America was an island continent during much of its history and mammalian evolution there went on in isolation for tens of millions of years. South America provides an unexpected natural but, in a sense, controlled experiment in evolution.

By the time I came on the scientific scene knowledge of the fossil mammals of South America had been carried considerably beyond where it was left at Ameghino's death. Activity in South America itself was somewhat limited, but Kraglievich and Cabrera, mentioned in the last chapter, had made important contributions and some younger men were starting to work, among them Paula Couto in Brazil, Rusconi in Argentina, and Hofstetter in Ecuador. Scott and Sinclair at Princeton University had published a series of large monographs dependent mainly on a collection made by Hatcher. Loomis, from Amherst College, went to Patagonia and published a book based on his collection sampling one fauna. Riggs, of the Field Museum in Chicago, had made large collections in Argentina, and Patterson was starting to study some of them. Some of the post-Ameghino studies were highly valuable and some were not — this is not the place to be more specific.

For reasons that I hope are becoming clear, study of the history of mammals in South America struck me as enormously interesting and of greatest value as a basis for the study of evolution in general. It also seemed to me that further collecting and study of the earliest faunas known to the Ameghinos and attempts to find still older mammals would be a good way to start. Moreover, after Carlos Ameghino and André Tournouër, no one except Riggs had collected fossils from the two earliest faunas (though Loomis had looked for some in vain) and no one was studying them. These were the reasons in back of my two expeditions to Patagonia in the 1930s. I have continued the study of South American fossil and recent mammals ever since and many of my technical publications well into the 1970s are devoted to them.

When we were in Patagonia in 1930–31 a number of geologists were working in the same region for the Yacimientos Petrolíferos Fiscales, the Argentinian government petroleum organization, generally known as the YPF (pronounced approximately ee-pay-effy). Two of these engineering geologists, A. Piatnitzky and J. Brandmayer, found some fragments of mammals at two localities in what they believed to be rocks of Cretaceous age, that is, of the last period of the Mesozoic, Age of Reptiles. They kindly guided me to those localities and our party made fairly extensive, although fragmentary, collections. These fragments were some of the most important discoveries of the expedition, although, as I have indicated, the initial lead was found by the YPF geologists and not by us.

By careful study both of the stratigraphy and of the fauna I was able to demonstrate that the YPF geologists were wrong in believing the age to be Cretaceous, but the fragments did represent the oldest mammalian fauna that had been found in South America. This fauna occurs in beds below those from which Carlos Ameghino obtained the oldest mammalian fauna known to him, called the *Notostylops* fauna by the Ameghinos but given its now usual name, Casamayoran, by Gaudry. I named the fauna and the beds in which the older fauna occurs Riochican because our first collection was made in Cañadón Hondo ("Deep Arroyo") which drains into the Río Chico ("Little River"), which is the outlet of the two central Patagonian lakes. Subsequently a richer and better preserved fauna of about the same age was found in fissures in a limestone quarry in Brazil, across the bay from Rio, and much of it was described by Carlos de Paula Couto, then at the Brazilian national museum in Rio. The exact age of the Riochican faunas has not yet been determined, but they are probably approximately late Paleocene (the first epoch of the Age of Mammals).

An important point that I set out to check both by field observation and by subsequent office study was the repeated statement by Florentino Ameghino that mammals of rather advanced type, such as occur elsewhere only in the Age of Mammals, were contemporaneous with dinosaurs in the Age of Reptiles in South America. That assertion had never been properly tested and therefore I had no reason to assume *a priori* that it was false. It needed to be tested thoroughly, however, because if true it would require profound changes in concepts of the history of life and of evolution in general. I found, in brief, that discoveries made by others than Carlos Ameghino did include dinosaurs but were either of unknown age or from rocks definitely older than any then known to contain mammals, including the newly discovered Riochican fauna. The fossils discovered by Carlos Ameghino were correctly reported as to relative age and some were associated with the oldest mammalian fauna found by him (our Casamayoran), but they were mere scraps and had been incorrectly identified by Florentino. They were not dinosaurs.

Mammals were already present in the Age of Reptiles. That fact was known well before the Ameghinos, and my own earliest important work was on those very oldest of mammals. They show an orderly evolutionary sequence of relatively primitive forms. The point that I tested and was able to establish was that Florentino Ameghino's claim of a gross contradiction of the orderliness of the sequence was incorrect.

Looking back over the last few pages I feel that I may have ceased to communicate adequately with some prospective readers. If I had spent (to be quite honest I would have to write "wasted") my life in breeding racehorses and one of them had won the Kentucky Derby and gone on to win also the next two races of the Triple Crown, millions of people would realize at once the significance and emotional impact of that outcome for me, and for an owner, a trainer, a jockey, and a good many other people. Many would even feel, without further statement, some participation in the event. I am somewhat at a loss as to how to convey the fact that such a thing as involvement in adding a whole new — and at the time, the oldest — fauna to the South American mammalian sequence was incomparably more thrilling and more important, not just to me but to everyone, than winning the Triple Crown.

It took a long time, but I did finish a two-volume monograph on the three oldest Patagonian mammalian faunas, those we now call Riochican, Casamayoran, and Mustersan. The first volume was published in 1948 and the second not until 1967. Plans to write a third volume mainly devoted to the geology of the region explored had to be dropped when I left the American Museum. I did continue to study South American mammals of all ages up to recent, partly in museums in Brazil and Argentina and partly in the United States with borrowed specimens from South America. In this connection I should mention not only the large museums in Buenos Aires and La Plata but also the rich but less diverse collections of the excellent relatively small museum in Mar del Plata (a very different city from La Plata). That museum started as the collection of an enthusiastic amateur paleontologist and was eventually taken over by the muncipality as a public institution. It has become an important research and exhibition center under the direction of Galileo Scaglia, son of the founder.

Perhaps the most interesting single one of my later studies of South American fossils was a monograph on a most peculiar group of mammals that appeared rather late in the history of mammals on that continent and disappeared without a trace soon thereafter, geologically speaking. The Ameghinos had some nondescript odds and ends of these animals and inevitably misunderstood the scraps, calling one *Argyrolagus* and some *Microtragulus*: *Argyrolagus* because they were found near the Río de la Plata ("River of Silver," *argyros* being Greek for "silver") and Florentino thought it was a hare (*lagos* is Greek for "hare"), and *Microtragulus* (Greek *mikros*, "small," and *tragulus*, Latinized Greek for "little goat") because Florentino thought it was related to *Tragulus*, not a goat but a tiny antlerless deer now living only in southeastern Asia and the East Indies.

A splendid series of specimens collected by the Mar del Plata museum and loaned to me by Scaglia showed that these animals had nothing to do with either hares or deer but were highly specialized marsupials, more nearly but still not very nearly related to opossums and kangaroos. They resembled kangaroos to the extent that they were bipedal, with short front legs and long hind legs and feet, leaping or ricocheting on their hind feet. Those pecularities were evolved independently in these South American animals and the kangaroos. The South American forms more closely resembled a bipedal Australian rat, more distantly related than even the kangaroos, and also our southwestern "kangaroo rats" (not kangaroos and not precisely rats although they are rodents). In fact this adaptive type has evolved independently over and over again in widely different places and from widely different ancestors. It is one of the most elaborate examples of the extent and also of the limitations of evolutionary convergence.

We still do not know much about the origins of the South American fauna. Only a few tantalizing scraps and not even a proper sample of a fauna have so far turned up from rocks older than our Riochican, and considerable evolution had already occurred within the fauna of the island continent by Riochican time. The earliest known South American mammalian faunas had extremely odd limitations. In North America, Europe, Asia, and, with high probability, in Africa, although the record there is not so good, the known mammals early in the Cenozoic (Age of Mammals) already represented a considerable number of basic, diverging stocks. In North America, for example, during the Paleocene, first epoch of the Cenozoic, there were already fifteen quite basically distinct known groups (orders) of mammals by a conservative classification. By a comparable classification, at about the end of the Paleocene in South America only seven such basically distinct groups are known.

There is more than that involved in this comparison and in the pecularity of the South American fauna right from its start. Five of the early South American groups were hoofed mammals, ungulates, which clearly had a common origin but had already diverged markedly by the end of the Paleocene. They could have, and quite likely did, all evolve in South America from one ancestral stock. Thus the early faunas there include derivatives from only three origins: marsupials (generally pouched mammals such as the opossums), edentates (eventually a mixed lot including armadillos, anteaters, and sloths), and ungulates (generally hoofed herbivores).

Those are the old-timers, and how such an unbalanced mammalian fauna with its limited repertory originated continues to be a great mystery. It was long supposed that all the forerunners were from North America. That such a peculiar small sample of the far richer North American fauna managed to go south would indicate that such spread had been by chance and across some strong barrier. That still may be the explanation, but it has become dubious, especially since possible direct ancestors of the South American edentates are not known from North America, although probable relatives are. Another, more recent hypothesis is that this was the basic fauna of Gondwanaland, a great southern continent before continental drift split it into Australia, Antarctica, South America, Africa, and India. But there is no evidence at all that any other of those regions ever had a mammalian fauna similar to that of early South America, and it is improbable that they did. The evidence is largely negative, and few scientists adopt a hypothesis for which there is no positive evidence. This hypothesis will become respectable only if or when ancestral edentates and ungulates are found in Australia and ancestral marsupials and edentates in Africa or India or, preferably, both. The chances do not look good. In any case, the mystery as to how such a peculiar faunal association ever arose remains, regardless of whether it once occurred more widely than in South America alone.

There were evidently many possible ways of life — ecological niches, in more technical terms — not early occupied by those old-timers. Their descendants did diverge and diversify into many niches, and they dominated the South American mammalian fauna for most of the Age of Mammals. Marsupials and edentates are still numerous but no longer dominant or so highly diversified in South America, and the direct descendants of the old-timer ungulates are entirely extinct.

Toward the middle of the Age of Mammals two new groups turned up in South America: rodents and primates. They quickly flourished, and are still prominent in the South American fauna. The continent was still an island, and the descendants of these newcomers evolved in complete isolation, soon becoming quite distinct from their relatives elsewhere in the world. They can only have come by island-hopping or somehow crossing a wide oceanic barrier against odds that prevented passage of their many faunal associates elsewhere. They must have come either from North America or from Africa, and we know of possible or at least conceivable ancestors for them from both these continents. That is a subject for much debate, most of it futile. The evidence at present is so evenly balanced that either side may well be right, but I incline slightly to the North American side, perhaps, as some of my French colleagues like to think, just because I am a North American myself. (I wish a North American, or for that matter a European, of African descent would go into this question.)

There is no room or reason for detail here, and the next big event came far along in the history of the age of mammals in South America, in its last few million years. What happened was that there was at first a trickle and then a flood of North American mammals spreading into South America and a smaller number of South American mammals reaching North America. North American invaders included ratlike and mouselike rodents, quite different from the old native South American rodents, swiftly diversifying and occupying the whole continent, although many of the older native rodents do survive. There were also carnivores, of the dog, bear, raccoon, weasel, and cat families. Hitherto the only mammalian predators in South America had been marsupials, evolved from oppossum ancestors. They became extinct just as the nonmarsupial (placental) predators were arriving from the north. A large number of North American ungulates also went south, and as they spread the old native South American ungulates all became extinct. So did some of the North American invaders, notably the horses and the mastodonts (the mammoths never made it), both of which also became extinct in North America. Tapirs, peccaries, and camels, also North American invaders, survive and are now more typical of South than of North America, and deer, also from North America, continue to be common on both continents.

Tropical Central America, previously North American both geographically and faunally, was environmentally more like northern South America and what survives there now is a mammalian fauna of mixed northern and southern origin more like the present also-mixed South American fauna. A number of South American natives pushed well beyond the tropics and into what is now the United States, but most of them became extinct. A noteworthy exception is the porcupine, a descendant of mid-Cenozoic South American stock. Less noteworthy, only because their natural distribution is still confined to nearly subtropical parts of the United States, are armadillos, descendants of the truly old-timers in South America. (We have only one species; South America still has some twenty.)

In broadest outline that is the grand history as I see it after nearly a half century of investigating it. It bears on and illuminates many of the problems and processes of evolution, ecology, zoogeography, and even such apparently remote topics as geophysics. This is autobiographical because delving into it and contributing to it has been so much a part of my life. What I have given here in so summary a way of course results from my own studies and conclusions. I have mentioned that doubt remains in several respects. Leaning over backward, I mention that a good authority on the living mammals of South America hardly believes a word of what I have said, and even invents words not really mine in order to disagree further. I will cite him in the notes to this chapter. Playing fair, I will also mention authorities on both fossil and recent mammals who do agree, in essentials, with all my conclusions. I think it is still fair to say that I know no one who agrees with my severest critic and quite a few who join in agreeing with me — bearing in mind that scientific matters are not settled by consensus, even though the consensus generally turns out to be nearer the truth than are its opponents.

Notes

§ William Berryman Scott was already one of the grand old men of paleontology when I entered the profession. He was forty-four years older than I and lived to be eighty-nine. He collected no fossils and did no field work after 1893, when he was only thirty-five. He visited Buenos Aires and La Plata in 1901 (the year before I was born) and was permitted by the Ameghinos to study and photograph specimens in their collections. He was never in Patagonia, and his voluminous Patagonian reports were based primarily on collections made by Hatcher. Under a misunderstanding, he was awarded a medal for exploration in Patagonia, and among his many amusing anecdotes was one about this recognition for exploration of a region without the usual prerequisite of having visited it.

§ Like Scott, Gaudry has sometimes been credited with work in Patagonia which in fact he never visited. A standard reference work says that he "investigated fossil animals in . . . Patagonia." I was only six years old when he died and of course never met him, but I have read and admired much of his work and I have a handsome portrait of him on a medal conferred by the Société Géologique de France. He scrupulously credited Tournouër with the actual work in Patagonia on which he reported. Except for Gaudry's work, now seldom read, and my paper on a part of the Tournouër Collection, poor Tournouër seems to have fallen into limbo. Loomis, who followed in some of his footsteps, did mention him but always misspelled his name as Tournier.

§ In the 1890s Osborn had planned to send an expedition to collect fossils in Patagonia, but an independent American Museum expedition there was not made until mine in 1930. However, in 1898–99 Barnum Brown went to Patagonia with Hatcher on the understanding that Brown's finds would go to the American Museum and Hatcher's to Princeton. All their finds were of a single fauna, later in age than any of the five that I collected, and all were available for Scott's studies. Osborn, who was my department head during both of my Patagonian expeditions, heartily approved my belated fulfillment of his hopes of the 1890s, but I must record that he did not help toward either planning or financing.

§ The Ameghinos named some fossil faunas and the rocks containing them by the names of a mammalian genus found in them. They did not know and hence had no name for the fauna, age, and rocks that I named Riochican. We now use modified geographic names consistently, and I have mentioned that Ameghino's *Notostylops* fauna (or *Notostylopense*) is now called Casamayoran. Following Gaudry, whose information came from Tournouër, this is named for Punta Casamayor, on the coast south of Comodoro Rivadavia. Carlos Ameghino recognized some rocks there as of *Notostylopense* age, but curiously enough no specimen in the Ameghino Collection can now be identified as surely from there; though specimens in the Paris collection, found by Tournouër under Carlos's guidance, can be. The next younger fauna was called *Astraponotense* by the Ameghinos, after *Astraponotus,* a genus that does occur in it but is very rare and is not a good index fossil. In 1930 Lucas Kraglievich proposed the name *Mustersense,* and this is now in general use (Mustersan in English). It is derived from Lake Musters, the western of the two central Patagonian lakes. (Musters was a nineteenth century British explorer of Patagonia.) Although the Ameghinos' Mustersan collection was mostly from south of Lake Colhué-Haupí, Roth collected Mustersan fossils (not recognized by him as such) from north of Lake Musters. I later studied both collections and my companions and I made further Mustersan collections both south of Colhué-Haupí and north of Musters.

§ The very oldest South American mammals now known by their own remains were found at a great elevation in the *altiplano* of Peru, raised by the Andean uplift. As yet they are represented only by a few scraps, more tantalizing than revealing. They lived near the end of the Mesozoic, the so-called Age of Reptiles. I noted "known by their own remains" because there are some footprints impressed in considerably older rocks in Patagonia that were possibly, but not surely, made by very primitive mammals. As they are not really identifiable, they add little to knowledge. The search goes on.

§   The relationships between the various marsupials, hares, and deer are those of ancestry and descent. Such relationships are fully analogous to your relationships with, say, a first cousin and a second cousin. They mean that you and your cousins had ancestors in common, closer in the case of first cousins and more distant for second cousins. In the evolution of organisms vastly greater durations of time and numbers of generations are involved. All mammals had a single ancestral group at a very remote time, but among mammals as they evolved further the common ancestry of hares and deer with marsupials lived earlier than the common ancestry of the various different kinds of marsupials. So my South American fossil marsupials were more nearly related to kangaroos, also marsupials, than to hares and deer, not marsupials. But the common ancestry of these South American marsupials and of the Australian kangaroos was not bipedal and was quite different from either one of those descendant groups. Their special resemblances, such as bipedalism, evolved separately, by convergent evolution, after they had split off from that remote common ancestry.

§   Some earnest students are likely to be among the readers of this sometimes unconventional autobiography, and I may throw a note or two their way. Here I will list the orders, as conceived quite conservatively, that I included in the Paleocene count for North America: Multituberculata, Marsupialia, Insectivora, Dermoptera, Primates, Tillodontia, Taeniodonta, Pholidota, Rodentia, Creodonta, Carnivora, Condylarthra, Pantodonta, Dinocerata, and Perissodactyla (only at the very end of the Paleocene and still a bit questionable there). In South America I included: Marsupialia, Edentata, Condylarthra, Litopterna, Notoungulata, Xenungulata, and Trigonostylopoidea. Less conservative classifiers might count up to twice as many North American orders and one or two more South American ones. Classification is an artifact, not an objective fact, and well intentioned efforts to make it fully objective without a share of art have so far failed quite dismally. Nevertheless the animals, fossil or recent, and their characters are objective and factual.

§   This chapter is quite serious and some readers may find it unduly technical. Yet the enjoyment of animals, recent or fossil, has an esthetic, emotional, and playful side. When I was first in Patagonia, still in my twenties, the playful side still sometimes took the form of doggerel, and I wrote a Patagonian bestiary which was not true poetry, to be sure, but like some now wholly antiquated poetry scanned and rhymed. It included a series of pieces, some now lost, the first devoted to a number of living animals, the others each to one animal, living or extinct. I had intended to give the first one here and let it go at that, but I could not resist some extension. With great self-control I give just one living and one extinct.

Living: The Guanaco

The guanaco is a camel but
    He hasn't got a hump.
He's about three-quarters mountain goat
    And seven-eighths a chump.
He is neither bold and dashing
    Nor reticent and shy
But if you get him cornered he
    Will spit right in your eye.
He has a disposition
    At which I only hint
And does a lot of things that I
    Don't dare to say in print.
The way he's put together shows
    A lack of proper taste
And one might say the food he eats
    Is only so much waste.
He dashes up the steepest cliffs
    As if it were good fun —
Which is really rather childish since
    He's nowhere when it's done.
He has a voice like nothing much —
    A rusty one at that —
And half the time it's off the key
    And half the time it's flat.
He's frowzy and he's lousy and
    His brains are simply numb.
Among our dumb companions he
    Is nearly the most dumb.

Fossil: The Entelonychia

A group of ancient ungulates so named by Florentino Ameghino. (In my journal for 13 January 1931 I inserted the German text for the defi-

nition of Entelonychia from Zittel's "Grundzüge der Paläontologie
(Paläozoologie), neubearbeitet von F. Broili und M. Schlosser." It starts
out boldly, "Ausgestorbene, plantigrade oder semidigitigrade planzen-
fresser mit vollständig, selten reduziertem Gebiss" and goes on and on
and on and on from there. I followed that by what I called a literal
translation from Zittel, and indeed it does convey, in its fashion, the
same diagnostic characters. I believe I noted before that the Patagonian
climate has odd effects on the sense of humor, as well as on other things.)

> Entelonychia I sing!
> The sweetest brutes that ever swum*
> Or walked on land or took to wing:
> *Homalodontotherium*
> > And *Notostylops, Colpodon*
> > — And thus I could go on and on.
> > (But you get the idea.)
> These beasts, alas! exist no more.
> O! burning shame that thus should pass
> A group that never bathed in gore
> But only dined on leaves and grass.
> > They walked quite firmly on their feet.
> > Their tooth-row always was complete.
> > (Well, nearly always.)
> We spoke of teeth: Then you must know
> Although a few had teeth before
> Enlarged, in general they go
> In gradual transition o'er.
> > Their gentle disposition shows
> > In just how small the canine grows.
> > (Almost like an incisor, in fact.)
> Their teeth were rooted and were plain,
> With no attempt to complicate
> Premolars, for they thought it vain
> More tooth-proud beasts to emulate.
> > In them such harmony one finds
> > That teeth were simple, like their minds.
> > (But more lophodont.)
> Their upper grinding teeth were wide
> And short — It is the better way.

*All right then, *you* think up a rhyme for *Homalodontotherium*.

The lower pattern does provide
Two crescent moons by night or day
  — But possibly you're tired of teeth.
  Let's have a look at what's beneath.
  (Their feet, I mean.)
Their carpal bones were alternate
And thus stood every stress or strain,
While I am very pleased to state
That as to hoofs their tastes were plain:
  A buxom clawlike hoof would do,
  But they'd accept the flat type too.
  (And the bone was notched at the end.)
Five toes they had and who'd want more?
Five toes, and who would do with less?
Entelonychids! Beast of yore!
Entelonychids! Heaven bless
  Your memory and may you be
  Examples for the likes of me.
  (And *you*, and *you*!)

References

Ameghino, F. 1906. Les formations sédimentaires du Crétacé Supérieur et du Tertiaire de Patagonie avec un parallèle entre leurs faunes mammalogiques et celles de l'ancien continent. *Anales del Museo Nacional de Buenos Aires,* 15 (3rd ser., vol. 8): 1-568. Ameghino published nothing in English and none of his important work in Spanish or French has been translated. This large, polemic volume is the best single expression of his ideas and so is cited in spite of the linguistic problem for some readers.

Hershkovitz, P. 1972. The recent mammals of the Neotropical Region: a zoogeographic and ecological review. In *Evolution, mammals, and southern continents,* ed. A. Keast, F. C. Erk, and B. Glass, pp. 311-431. Albany: State University of New York Press. This is the promised reference to a study by an authority on the living mammals of South America who believes that everything I have said and some things I have not said about their history are wrong. For another point of view in the same book see reference to Patterson and Pascual, below.

Loomis, F. B. 1913. *Hunting extinct mammals in the Patagonian pampas.* New York: Dodd, Mead and Co. A narrative of the Amherst expedition in 1911. Loomis also published not exactly a revision but a sort of summary of one of the Patagonian faunas in a companion volume.

Patterson, B., and R. Pascual. 1972. The fossil mammal fauna of South America. In *Evolution, mammals, and southern continents,* ed. A. Keast, F. C. Erk, and B. Glass. Albany: State University of New York Press, pp. 247-309. This is the promised reference to two of the authorities who believe that most of my views on the history of South American mammals are correct. For a dissent see reference to Hershkovitz, above.

Scott, W. B. 1937. *A history of land mammals in the Western Hemisphere.* Rev. ed., rewritten throughout. New York: Macmillan. A semipopular account including a review of the fossil mammals of South America excellent for its time and still useful. It has the peculiarity that it is written in what most paleontologists and indeed other readers consider the reverse order, from most recent to oldest — history backward.

———. 1939. *Some memories of a palaeontologist.* Princeton: Princeton University Press. A thorough, orthodox autobiography up to Scott's eightieth birthday in 1938. He lived nine years longer.

Simpson, G. G. 1948, 1967. The beginning of the Age of Mammals in South America. *Bulletin of the American Museum of Natural History* 91: 1-232, and 137: 1-260. Both with many plates. This is my description of the three oldest South American faunas of the Age of Mammals. The first volume includes an account of the work of the Ameghinos and others in Patagonia and a portrait of Florentino Ameghino.

9

How Not to Go to Mongolia

"Más vale algo que nada."
"Donde una puerta se cierre, otra se abra." — Both remarks put
by Cervantes into the mouth of Don Quixote.

The Don's wise sayings that something is worth more than nothing and
that where one door closes another opens can be paraphrased as, "If
you can't do what you want, do something else." One of the most outstanding of a number of examples in my life follows.

Having described the first known Mesozoic mammal skulls with Will
Gregory and being advised by Walter Granger that more could almost
surely be found where he discovered those, I greatly desired to go to the
place then called Shabarakh Usu ("Muddy Water") in the Gobi Desert
of Mongolia. In 1934 I had small but still potentially sufficient funds
available for the attempt. I had long looked forward to this and had
made some preparations for it.

At the time (late 1920s, early 1930s) as far as I could learn no one in
the United States could either speak or read Mongolian, and the State
Department informed me that no native Mongolian had ever entered
the country. I saw no means of acquiring an interpreter, as the large
previous American Museum parties had, so I managed with considerable difficulty to get a few Mongolian dictionaries, grammars, and
texts. I achieved a minimal speaking and reading ability: the language
was then being written in its own alphabet, which dates from the time
of Chingis (a more nearly accurate transcription of "Genghis") but is
well adapted to modern Mongolian.

With some friends I devised and played a game called Going to Mongolia, one of those games with markers advancing from square to

square by the throw of dice, with some squares obstructive and others helpful. In my version the winner arrived at Shabarakh Usu and found numerous Mesozoic mammals there. In the real world the obstructions were to prove complete.

In 1934, then, at the end of my second Patagonian expedition, I went from Buenos Aires to Paris, where I spent a delightful few days, and then on to Moscow, where I spent an unpleasant six weeks. It had become absolutely impossible to enter Mongolia by way of China, but there was no obvious reason why this could not be done through Russia, especially as there was a Mongolian embassy in Moscow. The United States was finally on nominally friendly terms with the U.S.S.R., and there was already an American Consul General in Moscow, Angus Ward, who happened also to be a student of Mongolia and its language and literature. Our first ambassador to the U.S.S.R., William Bullitt, arrived while I was there. With their backing I had official entrée into both the Russian bureaucracy and the Mongolian embassy, but it dawned on me little by little that this did not mean much, if anything.

The Russians said (surely with tongues in cheeks) that as Mongolia was completely independent only the Mongols could give me entry to their country. At the Mongolian embassy I had great difficulty learning anything, generally being told that only the ambassador in person could deal with this and that he was not there, even when I had seen him enter just before I did. Finally he condescended to tell me that of course only he could give me a Mongolian visa, but that as I would be crossing the Soviet (Siberian) border, I had to have U.S.S.R. permission before I could have a Mongolian visa. The Russians then told me again that they had absolutely nothing to do with Mongolian affairs but would cooperate if the Mongolian ambassador gave me a visa first. Thus I was batted back and forth for weeks without result, while I dangerously overstayed my visa to the U.S.S.R. itself. Finally Ward and Bullitt advised me to give up and get out, and I got, returning to Paris and eventually to the United States.

This maddening experience had some compensations. I became fairly well acquainted with Moscow, although I was not allowed into the Kremlin. (I also was not allowed to visit Leningrad or any place other than Moscow in the Soviet Union.) I quite enjoyed having both my telephone and my hotel room bugged, especially as this was done so ineptly that the telephone sometimes acted as a counter-bug and I could hear what was being reported by its listeners. I was charmed by a handsome female army parachutist (she said) who visited my room to see whether she could help me in any way, but I (somewhat reluctantly, I confess) declined her offers. I rather enjoyed an ex-American young communist once I had persuaded him that it was rather silly to follow me about when we could just as well walk and talk together. And for the first and alas! last time in my life I had all the caviar I could eat.

So, as I couldn't do what I wanted, I did something else, as will follow in due course. *Más vale algo que nada.* First, though, I must add two postcripts to my account of this fiasco.

One, the more important, is that what I could not do in 1934 the Poles have done more recently in conjunction with the Mongols and under the guidance of a very able Polish paleontologist, Zofia Kielan-Jaworowska. They went to Shabarakh Usu, now given the more propitious name Bayn Dzak ("Rich in Dzak," *dzak* being a Mongolian forage plant). There they found numerous Mesozoic mammals, just as I had hoped to, much better preserved than the original discoveries by Granger and his associates.

Somewhat parenthetically I might add that there have been a number of Mongols who disliked Russian control and have come to live in the United States. Among them was the Hutuktu, head of the Lamaist church (a Buddhist sect) in Mongolia and commonly although not accurately described as being considered a living god. He visited me in New York and we had a pleasant talk. I was delighted when my secretary told me that God was in the anteroom. (I regret to say that God needed a bath, not yet being quite Americanized.)

Also parenthetically I note that the Mongols no longer use their own alphabet but now write and print in the Russian (Cyrillic) alphabet, which is ill adapted to the completly different Mongolian language.

The second postscript is that the Russians, who with the exception of the parachutist were not really very hospitable in 1934, have several times in recent years invited me to visit Moscow and other Soviet cities at the expense of the U.S.S.R. Academy of Sciences. I have tentatively accepted more than once, but always difficulties have arisen and I have not again visited that country, although I do have cordial relationships with a number of Russian scientists by correspondence.

So that door closed, but others were already open, and more were to open within a few years.

My retreat from Moscow was late enough in the summer to make field work impractical that year, and I continued research mostly on subjects already somewhat familiar. There was still much to do on my South American fossils and from notes made in other museums. I was working on the National Museum's collection of ancient (Paleocene) mammals from the Crazy Mountain Field, as already mentioned. Another, smaller but likewise important collection of Paleocene mammals had been made some years earlier in southwestern Colorado by Granger but had not yet been completely described, and I wrote a short monograph on that, issued in three parts. The summer of 1935 was spent with Silberling in the Crazy Mountain Field. Our most important new discovery was of a fauna still Paleocene in age but somewhat younger than the fossils previously known from that area. The summer of 1936 I again spent in the West, but moving about over the high plains and in the basins of the Rocky Mountains to familiarize myself with most of the deposits containing fossil mammals of the first two epochs of the Age of Mammals, Paleocene and Eocene.

I have not mentioned before that the name of the Paleocene, one of my favorite epochs, is absurd. It is Latinized and then Anglicized from two Greek words, *palaios,* "ancient," and *kainos,* "recent." Thus it means literally "ancient-recent." *Eocene* comes from Greek *eos* or *heos,* "dawn," and again *kainos:* hence it means more or less "dawn of the recent," which is sensible except that it seems odd that the dawn should come after the "recent" was already "ancient." What happened is that the term *Eocene* was proposed by the great English geologist Lyell in 1833 for the first epoch of the Cenozoic (the era of "recent animal life"). Later geologists believed that the earliest Cenozoic was older than Lyell's Eocene, and Schimper, a paleobotanist, proposed in 1874 that a pre-Eocene epoch be recognized and called *Paleocene* because it is more ancient than the Eocene.

Schimper's Paleocene, based on fossil plants, considerably overlapped Lyell's Eocene, and there has been argument about that ever since. On the basis largely of fossil mammals a redefinition of the Paleocene was proposed in 1920 by the American W. D. Matthew (as I have mentioned before, he was technically a Canadian citizen but his college education and entire career were in the United States and I consider him an American). His definition is now almost universally accepted.

Let me digress and briefly go into the nomenclature of all the Cenozoic epochs in case some readers, like me, are etymophiles (that word, which I just made up, does not mean that they eat insects; it means that they love to know the origins of words.) Others have permission to skip.

The Cenozoic epochs, from young to old, are:

Recent or Holocene

Pleistocene

Pliocene
Miocene
Oligocene
Eocene
Paleocene

When Lyell started this out in 1833 his terms were:

Recent
Newer Pliocene (or Pleiocene)
Older Pliocene (or Pleiocene)
Miocene (Or Meiocene)
Eocene

His *Plio-* is from Greek *pleion*, "more," and *Mio-* from Greek *meion*, "fewer or less" the idea being that more species have survived from the Pliocene than from the Miocene. *Eocene* has already been explained. In 1839 Lyell proposed calling the *Newer Pliocene*, or *Pleiocene, Pleistocene*. Greek *pleistos* means "most," so the sequence "less recent," "more recent," and "most recent" is logical, although it is odd that "most recent" is less recent than *Recent*. Then in 1854 one E. Beyrich decided that the older part of the Miocene and younger part of the Eocene deserved recognition as a separate epoch, which he called *Oligocene*. *Oligo-* is from Greek *oligos*, "few", so the "few recent" (that is, with few survivors today) is older than the "fewer" recent. I have explained *Paleocene*, and I don't know who first used *Holocene* but its origin is obvious: Greek *holos*, "entire." I prefer *Recent* to *Holocene*, but remember that the geological Recent (capitalized) began some ten thousand years ago. The definitions of the various epochs have been changed and in fact most of the boundaries are still in some dispute as to detail, but the sequence is now universally used. Until quite recently the United States Geological Survey refused to recognize a Paleocene epoch, but lately they have decided to join the rest of the world's geologists.

Now the nonetymophiles can start reading again.

In 1937 I published two papers of no great importance in themselves but of biographical interest because they marked the opening of two doors. One was on superspecific variation in nature and in classification. Of course I had been classifying things consciously and technically as a graduate student and ever since, and I had been doing so unconsciously and nontechnically since I could talk if not before. (I can't remember when I couldn't talk and I have long had a conspicuously large vocabulary, but I have never been a compulsive talker.) All paleontologists classify things, and most of them have sense enough to know that organisms, living or fossil, vary and to take that into account. All along I had been thinking as I classified, becoming more and more concerned with the reasons, aims, and methods or in short the principles of classification. Now I had reached a point of sufficient experience and self-confidence to begin formalizing and publishing my ideas in this field.

The other opening door had a similar background but opened into a different field. The publication involved here was quite brief, only eleven pages with seven diagrams, and was titled "Patterns of Phyletic Evolution." I had long been intensely interested in and personally concerned with the theoretical aspects of evolution, again the principles involved, but I had felt that a contribution in that field should grow out of a broad factual basis in objective observation and should be accompanied by such a basic foundation. Now age thirty-five and more than ten years out of graduate school I dared to take modest personal steps in the direction of principles and theory. I was planning and working toward larger steps.

In many ways 1938 was a vintage year for me. To explain and introduce some of its happy events and outcomes, I must go back some years and mention as briefly and reticently as possible — but honestly — some highly unpleasant antecedents. After I left graduate school my first wife and I produced two more delightful daughters, making four: Helen, Patricia Gaylord (known as Gay), Joan, and Elizabeth. In spite of this bond, our marriage had become increasingly difficult. Finally in 1929 it became impossible and we separated, never to live together again. There followed years of excruciating efforts to face the finality of that separation, to protect those who were vulnerable, particularly the children, and to try to achieve some final legal and ethical conclusion to a continuously and impossibly painful situation for all concerned. All efforts failed until 1938, when I was finally given a hotly contested divorce. I put the children first in the care of their grandparents on both sides. Finally after my remarriage their custody was given to both my second wife and me.

When it became possible I married my childhood friend Anne Roe, whom I had met in Denver when she was two and I four years old, and who now in 1938 was a clinical research psychologist with a Ph.D. from Columbia University. That is the event that made the year 1938 an ecstatic one for me. I cannot express how happy our marriage has been and still is. She soon became the real mother of our children, who now have grown children of their own but continue to give us much pleasure. The only sorrow is that we greatly miss Gay, who died years ago shortly after her marriage and had no children. One of our grand-children also died, but we have seven living, all young adults and also a source of interest and pleasure.

My first mother-in-law and I remained on the best of terms even through my problems with her daughter and my eventual divorce and remarriage. My second mother-in-law, Edna Roe, was a really old friend, as we had met when I was four. While I was a child she was practically a second mother to me; if night overtook me at the Roe house I just stayed there. After I married her elder daughter she had some difficulty in recognizing that we were a separate family, although one where she was always welcome. "Mama," as I called her, remained a loved friend until her death at the age of ninety-three. She and my first mother-in-law became good friends and my first mother-in-law baffled acquaintances by introducing my second mother-in-law as "my son-in-law's mother-in-law." Anne has three sibs, also childhood acquaintances of mine, of course, and still highly valued friends. Bob, her older brother, is about my age and was my boyhood chum. Ed is Anne's younger brother and Pat (Mrs. Les Oldt) her younger sister.

The other great events of 1938 were that Anne and I finished the manuscript of a book on which we began collaboration even before our marriage and that we went off to Venezuela. I defer discussion of those subjects to the next chapter.

References

Matthew, W. D. 1920. Status and limits of the Paleocene. *Bulletin of the Geological Society of America,* 31: 221. This is only an abstract but it first defined the Paleocene essentially as now generally accepted.

Simpson, G. G. 1937. Patterns of phyletic evolution. *Bulletin of the Geological Society of America* 48: 303-314. This and the next entry are the two door-opening papers mentioned in this chapter.

——. 1937. Super-specific variation in nature and in classification: from the viewpoint of paleontology. *American Naturalist,* 71: 236-267.

Wilmarth, M. G. 1925. *The geologic time classification of the United States Geological Survey compared with other classifications accompanied by the original definitions of era, period and epoch terms.* U.S. Geological Survey, Bulletin 769. Although the U.S. Geological Survey has since changed a number of its own usages, there is no other convenient compilation of the original definitions of many of the terms still in general use.

10

Two Books and One Odyssey

"Good fun," he used to say, *"and every bit as exciting as algebra." –*
Norman Douglas

The character from *South Wind* quoted in the epigraph to this chapter
was speaking of languages, and I agree that both languages and mathematics are exciting and good fun. The events of 1938–39 included fun
of both sorts.

In the 1930s few biologists and especially few systematists were adept
in statistics. Most zoologists used no numerical methods more sophisticated than simple linear measurements of specimens and perhaps an
occasional ratio or average. If they were at all aware of more advanced
biometric methods such of those of Karl Pearson or Udney Yule in
England, they tended more to scorn than to use them. On the other
hand, psychologists were increasingly making their science quantitative
and statistical, seeking to counteract its earlier association with philosophy, which had given it the image of not quite a real science.

So it happened that although I was good at mathematics in college I
had never so much as heard the word "statistics" in my mathematics
courses, not even to mention courses in the natural sciences. Anne, on
the other hand, had learned basic statistical methods as essential in psychological research. I knew a good deal about zoology but very little
about statistics. Anne knew a good deal about statistics but very little
about zoology. This complementary relationship suggested that development of the application of statistical and related quantitative methods
to zoological studies could be forwarded by pooling our knowledge, a

sort of figurative marriage of minds that began before and has continued ever since our marriage in the more literal or legal sense.

My own impulse toward the marriage of statistics, or more generally of quantitative methods, and zoology arose from increasing interest in methodology and theory in the study of evolution and of systematics, the opening doors mentioned in the last chapter. In brief, I had come to adopt the following viewpoints or basic principles:

Populations, not individuals, evolve.

Populations, not individuals, are also the proper subjects of systematics and of its subscience classification.

An essential feature of populations is that the individuals included differ to some extent — vary — in many or most of their characteristics.

The actual subject of zoological study at the most basic level is a sample, be it one specimen or many, drawn from a population.

Quantitative measurement of variation in a sample is needed, and from this an estimate of variation in the population from which the sample is drawn.

An estimate of the probability of closeness of the quantitative features of a sample and those of a population is needed.

Comparison of two populations requires an estimate of the probability that two samples are drawn from distinct populations.

The basic operations called for in the last three statements are statistical in nature. But while applications of statistics to demography and biometry generally demand very large samples, and while such samples might be desirable in zoology also and are increasingly available, many zoological samples were, and many still are, very small, even single specimens. We wanted to decide whether such small samples could be treated from a statistical point of view, and if so, how. We had to bear in mind that few zoologists at that time were able, and even fewer willing, to use methods involving much more than grade-school arithmetic.

The result of such considerations and of much work was a relatively simple but fairly comprehensive book on the application of quantitative, including statistical, methods to zoology. The text was completed in 1938 and published in 1939. The book was far from being a best-seller, but it did circulate widely and a revision of it published in 1960 still sells modestly in spite of the fact that there are now several later and more extensive works on its subject. I believe that the original publication gave impetus to a revolution in zoological methodology if it did not quite originate that movement. Now it can be assumed that a properly trained zoologist is reasonably sophisticated in mathematics and especially in various statistical procedures. The invention and present wide availability of capacious electronic computers and pocket calculators have greatly increased the usefulness and the use of increasingly sophisticated methods.

Anne and I were married on 27 May 1938 and had a brief honeymoon driving down the Blue Ridge and to Charleston before returning to work. On 2 September 1938 we and my father sailed for Venezuela. Some fossil mammals had been discovered in the state of Lara and the equivalent of the geological survey (Servicio Técnico de Minería y Geología) of the Department of Development (Ministerio de Fomento) had invited me to carry on the investigation. They generously included both my wife and my father in the invitation, my wife for the duration, which turned out to be about eight months, and my father for a brief vacation. I was pleased for all of us but perhaps most especially for my father. He had long wanted to visit South America and had even learned some Spanish on his own. (He could recite the whole story of Goldilocks and the Three Bears in that language, an unusual accomplishment!) Now at last he had the opportunity.

In Caracas we became acquainted with the four principal geologists of the survey, with some of whom we would later share fascinating adventures: Santiago Aguerrevere, Victor López, Guillermo Zuloaga, and Manuel Tello. During various stays in Caracas we also became acquainted with most of the scientists and other intellectuals there, we became particularly good friends with William H. Phelps, Sr., a retired businessman of American origin but long a resident and citizen of Venezuela. He was an ornithologist, technically an amateur but of professional quality, who often visited the American Museum in New York, where we had first met. Now we became better acquainted and were sometimes guests of his wife Mona and him in his luxurious house. One of his sons, Billy junior, followed in his father's ornithological footsteps, and also became a good friend of ours. Related to but not part of the university there was a newly founded Institute of Geology with seven professors, two of them Americans and only two of Hispanic origin, including, interestingly enough, the professor of English and of mining and petroleum law. Only the two Americans (Newton Knox and Eli Mencher) were actually teaching geology.

Preparations to go out into the field proceeded at a leisurely, Latin American pace and we did not leave Caracas until 20 September, when we drove, or more exactly were driven, to Barquisimeto, a town of some size, the capital of Lara. There we met a Salesian brother, a teacher at the local (Roman Catholic, of course) school, Hermano Nectario María. For some years he had roamed the countryside, finding a number of fossils, and recently he had had a small grant for that purpose from the Ministry of Development. It was his discoveries that the Ministry had invited me to investigate, and he might have been jealous or perturbed but never gave any such indication. I will quote from my journal a description I wrote early in our acquaintance.

"Brother Nectario is a small, intensely active man with a quizzical, even whimsical, round face and snapping eyes. He is about the least restful person I have ever met, always hopping about like a flea and never finishing one thing before he starts the next. Every remark is an oration and a drama. To indicate that someone lives on a hill he cries 'Arriba! Arriba! Arr-r-r-r-i- - -ba!' with rising accent until he is screaming shrilly and with a rolling of the *r*'s like machine-gun fire. He pushes the auto through a bad place by throwing his body forward, shooting out an arm, and yelling 'p-r-r-r-r-t!' A stroll with him is a series of hops and jumps interrupted with 'Mire! Mire! Mire! Doctor! Le voy a dar una explicación!' Followed by a very positive (and often erroneous) graphic and explosive exposition of his ideas of local geology."

I trust that it is clear that I wrote not in ridicule but with genuine affection for another one-of-a-kind. Over the following weeks we became good friends, and we parted sorrowfully.

We spent the next few days staying at a hotel of sorts in Barquisimeto and driving with the Good Brother (as we came to call him between ourselves) to localities where he had found fossils. It was not until 26 September that we set up camp. The site selected for our camp was a ridge top, surrounded by ravines in several of which the Good Brother had found fossils. The ridge rose above a small village called San Miguel, at that time, and for all I know still now, inaccessible by motor. Our route was up the beds of normally dry ravines on the side of the ridge opposite to San Miguel, up to the top to a hut a mile and a half from our campsite. Although the ridge seemed deserted except for the one hut where we stopped, some fifteen men and uncounted women and children turned up and all our equipment and supplies were carried by hand to the campsite. It was too late to set things up, and it was two days before we actually moved in. We returned temporarily to Barquisimeto and while there said good-bye to my father, who was off to Puerto Cabello to catch a ship and return to the United States.

Our camp, home for a rather long time, consisted of an enormous fly, under which was a small umbrella tent, strictly for sleeping, with two cots, storage space and working and living equipment. A few yards away was a tent for cook and kitchen. The Good Brother had provided us with a foreman and cook, brother and sister, from Barquisimeto, but they proved to be unsatisfactory and were eventually replaced by local people who lived along our hill. In fact, with one other later, distant, and quite different exception, the brother-sister pair were the only Venezuelans to whom we ever took a dislike, perhaps unmerited. I later wrote, referring to Venezuelans of the poor or peón class, such as our servants and laborers, that on the average they could be "maddeningly vague and confused but pleasingly cheerful and friendly." It is true that we did occasionally become a bit querulous about friendliness that went to some excess. In that camp we were on view for a friendly audience even when we wanted to retire behind a bush for functions we considered wholly private.

After a shakedown period our staff was:

Capitaz (foreman): Trino López
Peones (our regular labor staff; three or four others were employed from time to time):
  Juan Raga
  José López
  Baudilio Giménez, Ximénez, or Jiménez (He thought it might be spelled any of the three ways, and in fact could not write it.)
  Sebastián Yepes

Cook: María Sánchez
Hunter: Rafael (I seem to have no record of his surname.)
Water and firewood boy: Pedro Regalado
Burro: Anonymous

The employment of a hunter, in addition to me, a bone-hunter, requires some explanation. Anne had anticipated that she would have little to do in her own profession while we were in Venezuela and therefore had undertaken to collect Recent mammals for the American Museum. At the museum she had learned the field procedures, and we had brought along suitable equipment. (A bag of tow, for stuffing mammal skins of medium size, was used as a maternity ward by a tarantula, who developed a large brood of young while we were in the camp near San Miguel.) Anne trapped small mammals, mostly rodents, near camp and made up the skins and skulls while I was prospecting or quarrying. Rafael ranged farther afield, as far as the rain forest visible from our desertlife hilltop, hunting for larger creatures.

Trino was a neurotic and quite puritanical man, but a thoroughly dependable, hard worker. María was neither neurotic nor puritanical, indeed rather the opposite. This difference in temperaments led to some problems, especially when Trino suspected, not without reason, that María was having an affair with Rafael. Few of the people on the hill were legally married, but most of them agreed with Trino that their substitute for marriage morally bound them to one mate for life. Fortunately, this quarrel blew over, as we had no intention of firing either of those useful and generally agreeable persons.

After extensive prospecting in most of the *quebradas* (ravines) round about, I found a rich-looking deposit of fossil bones near the trail down to San Miguel and decided to make our major effort quarrying them out. The men turned to with a will, pick and shovel in the hot and nearly breezeless quebrada. Some of them even learned how, with patience and some burlap and plaster, bones could be removed entire and not pried out in small fragments with a pick or machete — the Good Brother's system, partly from impatience and partly from an inevitable lack of professional experience.

Another attribute of Venezuelans was, and surely still is, that many of them have a sort of jocular attitude and personal humor that I find less common in other Latins. Our staff, with the partial exception of Trino, shared that agreeable trait. I should mention at this point that our people, like most Venezuelans of their class, were Spanish-Americans only by speech or courtesy but by heredity were mostly a successful blend of Indian and Negro, with perhaps just a slight dash of European Spanish. Sebastián was the comedian in our crew and generally kept all in good humor by constant monologue. A fair example:

Sebastián: Poor me. I am an orphan. I have to gain my bread with a pick in the sun. I am orphaned of my mother. I am orphaned of my father. (He was a man about fifty years old, a considerable age in that environment.) I am even orphaned of my godfather and my godmother.

Interlocutor: Hey! So you are a savage who was never confirmed.

Sebastián: Oh, yes. I am a good Christian. I was confirmed. Oh, yes. I was confirmed. I was . . .

Interlocutor: Well then, you must have a godfather.

Sebastián: No. No godfather. I was even confirmed without a god-

father. Poor me. I am a little orphan. I have no godfather. I have
no godmother.

And so on and on, all the while lustily making the dirt fly.

Things were going along quite well when one day the rains came,
and thereafter they went on and on. Venezuela as a whole has a fairly
well defined dry season, called "summer" (*verano*), and rainy season,
called "winter" (*invierno*) — "winter" is no less warm than "summer."
Our visit had been planned for the dry season, and we were taken by
surprise when a real rainy season set in. The area we were in is situated
in a sort of pocket at the end of a spur from the Andes extending north-
eastward through Colombia and eastward into northern Venezuela. We
now learned that here there is a regular rainy season, mainly in October
and November, when most of the country is in a dry season.

It rained and rained and rained. My quarry became a quagmire and it
was all we could do to save most of the bones that had already been
exposed. Anne's mammal skins mildewed and had to be kept dried out
as far as possible over a fire in the cook tent. Even our toothbrushes
mildewed, and that seemed about the last straw. There was little we
could do but sit under our fly reading and writing. Reading matter was
exiguous, consisting mostly of women's magazines sent by Anne's
mother. I took a permanent dislike to their fiction, which at that time
was largely concerned with unconsummated marriages. I also read a
German textbook on mammalogy aloud in English to Anne, who has
some German but does not like it and did not know the technical mam-
malogical terms. Among still other ways of passing the time, we wrote
alternate chapters of a murder mystery.

Another of the local climatic peculiarities was that in spite of its per-
sonalized rainy season our hill had an arid climate and desert vegetation,
notably the tall columnar cactus known as cardón, while the bottom
and the other side of the San Miguel valley had a more normal tropical
humid climate. There, before the almost impenetrable rain forest was
reached, there were several haciendas, comparatively prosperous, where
comparatively cultivated white Venezuelans lived in comparative luxury. On necessarily idle days we visited back and forth and were on
excellent and enjoyable terms with the haciendaros. They raised most
of their own food and as cash crops had coffee and sugar. We obtained
coffee beans direct from the tree and María took it from there extracting from them quite a delicious beverage. The sugar cane was transformed on the hacienda by primitive and highly insanitary methods to
dark brown cones of sugar known as *papelón*. Speaking of sanitation,

our only source of water (other than rain) on the hill was a puddle not
only muddy but also insanitary for various obvious reasons. The mis-
tress of the nearest hacienda kindly offered to keep us supplied, via
Pedro and Anonymous, the burro, with distilled water (*agua destilada*).
We accepted with delight and drank it even after we discovered that in
the local dialect *destilada* does not mean "distilled" in the English sense
of the word but that the water has been filtered by oozing through the
sides and bottom of an unglazed coarse earthenware pot, which does
remove much of the mud but none of the germs.

There were many small events during this time, fully chronicled in
our journal but perforce not here. In spite of the rain we did get in to
Barquisimeto, with difficulty and sometimes peril, a couple of times.
There was a fiesta in San Miguel. There was a total eclipse of the moon.
The next day, ever avid for folklore, I asked whether anyone had
noticed anything peculiar the night before and was told, with indifference "Oh, yes. There was an eclipse." There was a threat of revolution, but it didn't come off. A stranger turned up in camp and after
much questioning and evasion someone said he was María's husband.
"But we understood that María is not married." "That's right. She
isn't." In that society you are not necessarily married to your spouse.

We went to Caracas for the holidays, which we spent delightfully
with the Phelpses, and to consult a doctor, as I had been rather ill, perhaps from drinking *agua destilada* but probably not. We went back to
the San Miguel camp after the holidays, and finally broke camp and left
on 26 January. A few days were spent in Barquisimeto, mostly packing
fossils, which, incidentally, were finally shipped to New York in solid
mahogany boxes. (One of the laboratory men, a good household carpenter, made elegant furniture out of the boards after the fossils were
unpacked.) We left the state of Lara, not quite for the last time, and
went back to Caracas on 5 February 1939.

Before ending this (to me) all too brief account of our stay in Lara I
must record one more of many conversations among the workmen and
add a brief note on their way of life.

At work in the quarry, someone mentioned African savages:

Sebastián (dark coffee-colored): Well, we can't talk. We're of that
same race ourselves.
Trino (chocolate-colored): Certainly not! We are Christians and
white men.
Juan (tan): *Some* of us (meaningfully) may not look exactly white,
but really we count among the white races.

Trino: If a man is civilized, like us, he is white. No one has a white skin like paper. Only savages are really black.
Sebastián (far the most negroid of the group): Well, of course I can't say what you are, but as for me, I am Dutch!
[General laughter.]

For all their good humor, willingness, and general decency, these men, their families, and all the dwellers along our hill were desperately poor and by most criteria miserable. They lived in stark mud huts without lights, water, furniture, stoves, or any sanitation. Women, although modest, dressed in a single garment, men, also modest, in two (shirt and trousers). Each family had a few chickens, and as soon as an egg was laid it was rushed, still warm, to us for sale. There were a few pigs, but meat was eaten only as a great treat on special occasions. The staff of life was Indian corn (maize) raised in small patches on the nearly barren soil where they squatted. They had no schooling at all and were almost completely ignorant of the world. One woman asked us whether the United States of North America, whence we said we came, was as far away as Caracas and what state (of the United States of Venezuela) it was in. We were briefly a sort of relief program since we paid our men (in silver coins) at the government wage scale, which no local employer would have dreamt of doing. Before and after that windfall, they lived somehow and they enjoyed life, although usually not for very long. In Caracas we enjoyed the hospitality of a family of great wealth and visited with others. I leave the moral, if any, to the reader.

An absurd incident somehow comes to mind at this point. One morning in Barquisimeto, wearing shoes with silent rubber soles, I rounded a corner and suddenly came face to face with a donkey. Startled, he unseated his rider and with great clatter sat down himself in the middle of the street. Great confusion! Crowds gathering! "What goes on?" "What's got the burro?" "Oh, that man — see, that man there! My burro took one look at him and was scared to death!"

When we returned to Caracas the Survey geologists there were planning a grand expedition to the region called the Gran Sabana in southeastern Venezuela. They invited us to go along, and we gladly accepted. It was to be two or three weeks before the expedition would get off, however, and in the meantime we decided to visit the *llanos,* the large nearly flat region between the coastal northern mountains and the Orinoco River. Among other reasons, I wanted to investigate some reports that fossil animals had been found in that region.

On the way, not yet quite clear of the hills, we stopped in San Juan de los Morros, a resort town, often mentioned as a locality for mastodons, usually but wrongly called "mammoths" by the Venezuelans. I could find no record of any specimen that *surely* came from there, and after what I must confess was a somewhat desultory investigation on the spot I decided that in all probability none had, so on we went. In passing and in a moment of nostalgia I recall that a room and full board in the attractive resort hotel cost the equivalent of U.S. $3.80 per day. I was annoyed all over again that the not at all attractive hotel (in my journal it is a "stinking dump") in Barquisimeto charged just the same amount. At our next stop, in Valle de la Pascua, the hotel was even worse than that in Barquisimeto, but it charged only half as much.

On, as soon as possible, to Zaraza, a small town east of Valle de la Pascua in the northern llanos. Here we had some trouble finding quarters, but were taken in by the oriental houseboy of an absent American geologist and enjoyed relative luxury thanks to the absentee host and his factotum, who was also a good cook. There is a definite record of a quite interesting fossil mammal found near Zaraza, and I wanted to visit the locality, which was some distance from town.

I soon met another fairly extreme example of a one-of-a-kind. As my journal has it: "A gentleman on horseback; a shoe and a spur on his left foot, an *alpargata* (canvas shoe with rope sole) on the right; shirt-tail hanging out; carefully holding an umbrella, black cotton, over his head with the same hand that held the reins of the horse; no hat on but a fine shock of black hair; a shrewd, intelligent face, set off by a long tuft of chin-whiskers. This is Doctor Torrealba, a famous character in Venezuela."

He knew I was coming, knew about the fossil locality, had read the publication about it in German, knew a man who could lead us right to the spot, got hold of the man, Morales by name, and over the latter's delaying tactics bullied him into taking us there the next day. Next morning, only an hour and a half late, we took off as a cavalcade, literally so as we were on horseback or in some cases, including mine, muleback. We found and I thoroughly examined the fossil locality but there were no identifiable mammals. There was a large fossil turtle in a block weighing some two hundred pounds, and I decided to come back and take it out when I had more man power. A few days later a very rough track was cleared out with machetes, we brought our truck within a half mile of the fossil, and it was carried to the truck by a group of some eight local men. It did eventually get to New York for preparation and study and then back to Caracas as property of the Venezuelan government.

We visited Dr. Torrealba's home, which was not luxurious, to be tactful, and where there were two women and a large flock of children of uncertain status. He had, however, an excellent study-laboratory where he was doing highly competent research on Chagas' disease. That disease, endemic in the region and the cause of nearly universal debility and considerable mortality there, is caused by a trypanosome, a close relative of the cause of the dread sleeping sickness of Africa. Here in Venezuela the trypanosome infects a number of animals, and at the time of our visit the doctor was working on the identification of these vectors. The disease is carried from them to humans by a cone-nosed bug, which thrives in all the local huts. As a small return for the doctor's help to us, we undertook to have some of the suspected vectors precisely identified for him. That caused a little discomfort, as one of the specimens was a skunk inadequately preserved with formaldehyde, which became an unpleasant traveling companion on the hot days as we drove back to Caracas.

We were back in Caracas on 23 February, spent most of a week developing film and printing photographs, and flew to Kamarata in the Gran Sabana on 4 March. In the first chapter of this book I gave an account of one day, perhaps the most exciting among the many exciting days, while we were with that expedition. I also mentioned with utmost brevity some of the things we did at Kamarata. I do not now propose to expand much in spite of temptation — we still have far to go in this contemplation of my life.

I should record that Uruyén, near which the expedition's main base was set up in the Kamarata Valley, was not a village in any usual sense but just a couple of thatched huts full of Kamarakoto Indians. I spent my time learning and recording everything possible about the Indians with the effective help of one who had a smattering of Spanish. Anne spent most of her time collecting Recent mammals with the rather ineffective help of an Indian assistant. As a result of our work there I wrote a rather large book, all about the Indians and not a narrative of our experiences. Anne wrote a paper on the living mammals; she is probably the only psychologist who has in her bibliography a paper in Spanish about the living mammals of a remote region.

I must record, too, that while the Indians varied in character, as all people do when you come to know them, we found them quite charming. They are ethnically Caribs, a once far-flung group speaking languages at least as diverse as the various Romance languages in Europe. They were reputed very fierce and cannibalistic by the Spanish explorers and earliest settlers, perhaps as an excuse for slaughtering them. They were in fact annihilated in the West Indies and everywhere well-settled by the Spanish, and they now survive only in such remote refuges as the Gran Sabana of Venezuela. The Kamarakotos never had or have strangely lost the ferocity imputed to all Caribs by the early Spanish. Their language, of which I made the first record, is a dialect closely akin to but not identical with that of the Taurepanes of the region near Roroimá (or "Roraima") around the junction of former British Guiana (now Guyana), Brazil, and Venezuela.

I understand that tourists are now regularly flown into Kamarata. I shudder to think what this may have done — be doing — to the places and the inhabitants which and who were so nearly pristine when we were there.

Before leaving Venezuela we made one last excursion, this time by car into the high Venezuelan Andes, which loom above Lake Maracaibo. The scenery is grand, the going usually difficult, the towns attractive. On the way, going and coming, we stopped at Barquisimeto and again visited our friends and former employees near San Miguel. With Brother Nectario I made arrangements for some continuation of work there and further shipment of fossils, but these plans did not work out.

We returned to New York in May 1939.

Notes

§ *South Wind,* source of the epigraph for this chapter, is one of my favorite novels. It is almost as full of quotable bits as Pope, Shakespeare, or the Bible, those famous mines of quotations. I toyed with the idea of using only epigraphs from *South Wind* in this book, but decided on a greater variety of sources. Still I hate to waste all those I had in mind from *South Wind,* and so I give a few of them here:

> *Poets are a case of genepistasis.* (Try that on your dictionary; all of mine in English failed the test, but sense emerges if you consider the Greek roots.)

> *I find everything wonderful and nothing miraculous.* (That is an essential feature of my own religion.)

'Things are happening here,' he said. (I have seldom had a day when things were not happening.)

Mediaeval minds knew many truths hostile to one another. All truths are now seen to be interdependent. (One can go on to the implication that what is not interdependent is probably not true.)

§ Mention of the approach used in our joint study of quantitative zoology gives me a chance to complain of the abasement of what was originally a precise and needed technical term to the level of vague and unnecessary jargon. A *parameter* is a statistic of a population. What is actually calculated from measurements of a sample is an estimate of such a parameter. It has become more common than not to use the word *parameter* not for the (never exactly known) population statistic or for the estimate of such a statistic but for some variate such as the dimension or measurement involved, for example the length of a measurement in some specimens. I suppose that those who thus misuse the term think that *parameter* makes them sound more profound or more in-group than such common words as *dimension, measurement,* or *variate*.

§ The collection of fossil mammals made in the region of San Miguel was large but not very well preserved. Its preparation, performed in the American Museum fossil-vertebrate laboratory in New York, was particularly arduous. It revealed a fairly typical South American fauna of the late Pleistocene and added to knowledge of the distribution of such faunas. Giant ground sloths, old natives of South America, predominated and there were other old natives, but there were also invaders from North America, such as horses, deer, and wild cats. The most interesting discovery was a new genus of the old native hoofed mammals called toxodonts. A description of that was made and published by one of my graduate students.

Not long after our stay in Venezuela there was a complete change in government and the new ministry demanded that the whole collection be returned to Caracas even though study of it had not been completed. The collection was returned and receipt duly acknowledged, but a South American in an anti-*Norteamericano* mood afterward publicly claimed that I had stolen it. As far as I know it is still sitting largely unstudied somewhere in Caracas. This sad outcome had nothing to do with anyone named in the present account.

§ The fossil turtle collected near Zaraza turned out to be an interesting new species, which I named *Podocnemis geologorum. Geologorum* ("of the geologists") was a tribute to the four principal geologists then operating the Venezuelan geological survey.

§ My monograph on the Kamarakotos was written in English but translated in Caracas and published in Spanish as cited below. The English version, elaborately coded for subject indexing, was reproduced in 1969 by the Human Relations Area Files at Yale University and thus became more readily accessible to anthropologists. Some of my best friends are anthropologists, and I have published quite a few other things in the field of anthropology, broadly defined, but anthropologists have not paid much attention to this one. I think the reason is that its contribution of new material was almost all objective description without much theoretical, philosophical, or even methodological predilection such as is now, for better or for worse, more in the mainstream of anthropological thought.

§ My remark about the reputed former ferocity of the Caribs and the present good nature of the Carib Kamarakotos reminds me of an example in which a people's ethos has certainly changed in only a generation or two. Nineteenth-century Fijians were beyond any doubt extremely pugnacious, habitual cannibals. Their twentieth-century grandchildren are charming, gracious people who never eat either each other or missionaries and seldom kill them.

References

Simpson, G. G. 1940. *Los Indios Kamarakotos (Tribu Caribe de la Guayana Venezolana).* Caracas: Ministerio de Fomento. This is the Spanish published version of my monograph on my favorite Indians. The reference number for the coded English three-volume pamphlet version in the Human Relations Area Files at Yale University is 5: Simpson, N-5, (1939) 1940.

Simpson, G. G., and A. Roe. 1939. *Quantitative zoology: numerical concepts and methods in the study of recent and fossil animals.* New York: McGraw-Hill. A revised and rewritten version by Simpson, Roe, and R. C. Lewontin was published in 1960 by Harcourt Brace Jovanovich, New York.

Sokal, R. R., and F. J. Rohlf. 1969. *Biometry: the principles and practice of statistics in biological research.* San Francisco: W. H. Freeman and Co. I cite this as a good example of the more complex works now available in the same field as Simpson and Roe 1939.

11

An Olio of Things Neglected But Not Forgotten

Sed summa sequar fastigia rerum [which might be roughly translated as, "But I shall hit the high spots."] — P. Vergilius M.

At this point I feel something akin to panic. Insofar as this account is chronological it has brought me to the age of thirty-seven, which happens to be just half of my life up to the time at which I am writing. At thirty-seven my character, such as it is, was maturely and firmly established. I was also well established in my profession, with one of the best jobs then available in it and already a member of the oldest and one of the most prestigious American learned societies. Yet in looking back over the preceding chapters I see that I have slighted or even omitted important things that went into the forming of my character and that influenced my choice of profession and progress in it.

Of my parents I have said little more than that my mother disliked her in-laws and my father finally was able to visit South America, yet they obviously greatly affected my character and progress toward a profession. I have said nothing about my brief career in business, which turned me back to science as a profession. I have said nothing about early wanderings and have made only the most cursory reference to my love of nature and of mountains, deserts, and seas. I seem almost to have shied away from some of the things to me most dear, or most formative, or both. I will now try to catch up a bit on a few of these neglected topics.

Parents

Diligere parentes prima naturae lex est — Valerius Maximus

Many children do not agree with Valerius that it is the first law of nature to esteem one's parents, but I did and I still revere their memories.

I was an unexpected last child, and being different from other members of the family or indeed any known relatives both in temperament and in appearance, I was something like a cuckoo in the nest. It seems symbolic that unlike any ancestors (or, now, descendants) I had bright red hair. (What little remains is now pure white.) My older sister thought it both clever and hilarious to reply to the question "Are you ready?" "No, but I'm Reddy's sister."

Except in one respect, to be specified in a moment, my mother simply loved me and overlooked or even admired my peculiarities. My father, however, felt that he should do something about them, and at times when I was young this led to some difficulties. I was a sickly child but I did play often and vigorously with neighborhood children, especially the Roe kids. Nevertheless, my father, although highly literate himself, feared that I was too much so, and he regretted that I was not more athletic. One of the things he tried to do about it was to play catch with me. I was hit by the ball more often than I caught it, and it took a long time to persuade my father that because of my peculiar eyesight it was literally impossible for me to keep my eyes on the moving ball. In other ways, too, he occasionally tried to push me beyond my physical capacity. But almost always he treated me as a good friend.

On moving to Colorado my father was for a time a claim adjuster for the Colorado and Southern Railroad and then went into irrigation and land development, especially along the Green River in northeastern Utah. He often took me along when he visited that project or went to Salt Lake City on business connected with it. When I was a bit older we frequently went for weekends or longer holidays in the Colorado mountains; each with cans of food rolled in a blanket tied at the ends and slung over a shoulder, we would walk all day at a leisurely pace and sleep on the ground whenever night overtook us. That was completely delightful and I acquired a deep love for nature, and particularly mountains, that has stayed with me all of my life and that was a factor in my entering a profession that included camping out during many active years.

My father went from land development to mining and for a time operated a mine, primarily for gold, up Buckskin Gulch a few miles from the village of Alma in South Park, Colorado. In my teens I spent two summers working there. One summer my chum Bob Roe did also, and the whole Roe family visited. The site is magnificent, a glaciated valley with conifer forest in its lowest stretches but rising well above timberline. The mine and mill were surrounded on three sides by mountains all over 14,000 feet high — in one single day my father, Bob, my sister Martha, and I climbed to the summits of four of them. The mine and the mountains established in me a bent for geology.

I was my mother's favorite and she indulged me in almost all ways. The exception was her ironclad rule that my sisters and to only slightly lesser degree all females were sacrosanct. If I got into a fight with one of my sisters, which happened often with the elder of them, I was automatically at fault no matter how the fight arose, and I was turned over to my father for punishment — a system that I considered (and still consider) unfair both to me and to him. It was at first difficult for me in later years to realize that women are just as human as men, and by and large more fun in other ways.

With few exceptions my parents were both supportive and permissive. Even when quite young I was usually allowed to go my own way, which was always my tendency, and I was loved and helped if on occasion that brought me to grief. Throughout their lives and far along in mine that attitude continued. My mother lived to ninety, my father to eighty.

My Career in Business

"If everybody minded their own business," the Duchess said in a hoarse growl, *"the world would go round a great deal faster than it does."* — Lewis Carroll

My first engagement in business occurred when Bob Roe and I built a stand at the corner of Milwaukee and Eleventh streets in Denver and stocked it with bottles of soda pop on ice in a tub. We bought the pop by the case wholesale, got ice (probably at a discount) from the man who serviced our family ice boxes, and used a washtub of one of our mothers. We soon built up a regular clientele and made a pretty penny selling bottles of pop at five cents each for their contents only; we kept the bottles to turn in, of course.

Thereafter I worked whenever and at whatever I could through high school and into the college years, sometimes at odd jobs and sometimes at peculiar ones. At various times I was a key boy at a resort swimming pool; a combination bellhop, electrician, and engineer at a hotel in another resort; a house painter (it was a courthouse that I painted); a trail maker for the Forest Service; and a guide in Estes, or more properly Rocky Mountain National Park; among other things. The most serious job in my early teens was with a map company while I was out of school for some months following one of many illnesses. I made blueprints and delivered them by bicycle. I worked about sixty hours a week and was paid four dollars; two dollars of which I gave to my parents for board (they were broke, too) and two dollars of which I put in a savings account in a bank that later went out of business without paying back any of my savings.

All those jobs were just to pass time and help out. After my freshman year in college family finances were at the lowest ebb ever, and I dropped out and went to Chicago, where my half-uncle proposed to start me on the way to a career that might eventually be as lucrative as his. He was on the Board of Trade, which despite its misleading name is an exchange where speculators trade in grain and some other commodities. My uncle got me a job as one of the boys who ran orders from desk to pit during open hours and stuffed envelopes with market reports after hours. It was boring, and besides, I decided in my overly straitlaced way that it wasn't honest. I didn't see why men like my half-uncle, whom I considered merely a gambler, should make fortunes trading in grain they hadn't raised, didn't deliver, and in fact never saw and never did anything with or for. (There really is an almost honest rationale for this activity, but I could never have been happy in it.)

I quit the Board of Trade and got myself a job at the Cable Piano Company, also in Chicago. I started as a sort of office boy in the advertising department, but I was soon given a desk job as sales manager for books entitled *The Hundred and One Best Songs* and *The Hundred and One Best Poems,* published by the advertising department of the company. I was seventeen years old and I had a private secretary at least twice as old and at least ten times as experienced as I. She was not as pleased with the arrangement as I was. Come spring, I was bored with this job and fed up with everything I observed about the advertising business and the piano business as well. I quit. I vowed I would never become a businessman, and I never did.

I set out to see the world, starting by walking to New Orleans. I worked one last deal in publicity: I persuaded a manufacturer of shoe lining to pay me to walk with one shoe lined with his product and the other with an inferior product that we counted on to wear out sooner.

In fact they both wore out at the same time near Cairo, Illinois, and I went on by freight (unknown to the railroad) from there. The shoe lining manufacturer was decent enough to send the agreed pay to New Orleans in spite of the publicity failure, and after a quite episodic stay in Louisiana and Texas I went back to Colorado and eventually back to school.

Mountains, Deserts, and Seas

To me high mountains are a feeling, but the hum of human cities torture. . . .
O that the desert were my dwelling place! . . . Roll on, thou deep and
dark-blue Ocean, roll! — Byron, *Childe Harold*

From a second-floor porch where I slept during much of my boyhood in Denver there was a magnificent view of the Front Range of the Rockies that delighted me every morning as I woke up. The view is dominated by three 14,000-foot peaks: to the left, far south, Pikes Peak; straight ahead, to the west, Mount Evans; to the right, far north, Longs Peak. I have mentioned climbing several 14,000-footers with my father, and I climbed Mount Evans with him just before I left to live for many years in eastern United States, not long before that mountain was, from my point of view, descecrated by the building of a road to its summit. I also climbed, not with my father, Longs Peak, which to this day has no road and I hope never will. We scorned Pikes Peak because even when I was a child it had both a road and a cog railroad to its summit, making it unworthy of being ascended on foot.

I climbed other 14,000-foot peaks — they are abundant in the Colorado Rockies — and some less high, although it used to be difficult for me to consider anything below 14,000 feet a real mountain. I often climbed alone, which no one should ever do, and I even pioneered some new routes alone, which on percentages should have been the death of me but, as you see, was not. I used to say that I had never been on a rope, but a friend who had been in college with me sent me a photograph showing me on a rope on the Arapahoe Peaks, a climb that I had forgotten. (Those are good, rugged peaks, but not among the highest ones.) When I was factotum at a resort hotel — that was at Grand Lake Lodge, then new — I used to take Sundays off to go climbing by myself. That so infuriated the manager, who expected the cheap help to be on call twenty-four hours a day and seven days a week, that he finally ran me out at the point of a gun. He still owes me some pay.

Since leaving Colorado I have seen, but have not climbed, the highest mountains of every continent except Asia. I still hope someday to get a glimpse, at least, of Mount Everest. Australia, the low continent, has no really spectacular peaks, but on the other continents Mount McKinley, Aconcagua, Monte Rosa, Kilimanjaro, and the peaks of the Antarctic Peninsula and Victoria Land are all spectacular enough. To a person with my tastes, and with the reservation that I have not seen the Himalayas, the most wonderful scenery on earth is that of the Antarctic Peninsula, rugged, not ice-covered but with just the right amount of glaciers and bergs — all that and penguins too! (I'll add something about penguins later.)

A nurse here in Tucson, where I now live and am writing this, once remarked to me that she was leaving. She couldn't stand it. No matter where you look there were those horrible mountains. That is exactly what Anne and I like most: mountains in every direction. They are not high. Highest are the Santa Catalinas just north of us, rising only to somewhat over 9,000 feet. But they are all beautiful.

Waterfalls are among the main attractions of many mountains, although not of desert mountains, and although some of the most spectacular falls are not in mountains at all. I have noted that Anne and I were among the first to see Angel's Falls, highest on earth, which does drop from a mountain (a *tepui*). Also in South America, but not in mountains, is Iguaçu, in another sense the largest of all falls. Victoria Falls, between Zambia and Rhodesia (which may become Zimbabwe almost any time now), merits its fame, and so, somewhat to our surprise, does the Gullfoss, in Iceland, despite being comparatively small.

Another attraction of Tucson is that it is in a fine desert. Coming in by plane once I overheard some transit passengers remarking that they never stayed here because there is nothing but sagebrush. In fact there is no sagebrush at all in our desert, the Sonora Desert, but there are hundreds of other kinds of plants. Many of the perennials have spectacular flowers every year, such things as palo verde, ocotillo, yucca, and several kinds of cacti. When spring rains come at just the right time, as they do every few years, dormant seeds of annuals spring into life and floor the whole desert with an unimaginable display of varied colors. This desert supports a rich fauna, the more fascinating because it marks the northern limit of many birds and mammals rare or absent elsewhere in the United States, such as Inca doves, peccaries, and coatimundis. It also had human inhabitants long before white men invaded it, and their descendants are still abundant here — mostly Pápagos in our part of the desert.

The idea that the Sahara Desert is only a vast sea of sand is almost, but not quite, as silly as the idea that the Sonora Desert is only a sea of sagebrush. There are indeed vast stretches of sand in the Sahara, but so are there vast stretches of rocks, mountains, watercourses dry most of the time like our southwestern arroyos but running after rains, soil with sparse but widespread vegetation, and oases smaller in area but blooming and numerous across the breadth of the great desert. Here too, there were people long before invasions by Phoenicians, Romans, Arabs, Italians, French, and Spaniards. And here too the ancient peoples still are present and seem to me even more interesting than the geology, flora, and fauna or than the inhabitants of the Sonora Desert, perhaps because the Pápagos are more like home folks.

In most of North Africa west of Egypt the descendants of the prehistoric peoples are collectively known as Berbers. Along the coast and even into villages of Saharan oases they have become Arabized and are considered, even by themselves, as Arabs, because they are Moslems and speak modern Arabic dialects. Nevertheless, ethnic or tribal Berbers still occupy much of the vast extent of the Sahara. In Algeria, the only region where I have personally encountered them, some, the Kabyles, even persist almost at the edge of the large Arabized city of Algiers. In the Algerian Sahara there are numerous tribes, the best known of which are probably the Ouled Naïls and the Tuaregs.

Until recently, surely, and probably even now these tribal Berbers had customs grossly contrary to the tenets of Islam. The Ouled Naïl girls came (come?) into the oasis villages where they danced in the nude and practiced prostitution until they had accumulated an ample dowry in gold coins worn as necklaces. Then they returned to the tribe, where the most successful prostitutes were considered the most desirable and settled down as fruitful wives. It is amusing that the village musicians, faithful Moslems, turned their backs while playing for Naïl dances so that (in theory) they would not sin by looking at naked women. Nevertheless in one oasis a French painter, Etienne Dinet, converted to Islam, violated the dictates of his new religion and became famous as a painter of nude Naïliat.

The Tuaregs (or Touaregs — all these names have variant versions) live in the deep Sahara and have a curious twist of usual, or especially of Moslem, sex roles. The men wear veils; the women do not. The women are free in manner and speech. They often do the courting. In their mountain country inscriptions on many rocks proclaim feminine willingness and desires or give public notice of what would elsewhere be clandestine trysts. Until recently, at least, these inscriptions were done in a peculiar alphabet called *tifinagh*. In ancient times, well over a thousand years ago, the use of this alphabet for writing Berber languages stretched all across North Africa from Egypt to the Atlantic and into the Canary Islands. It was possibly derived from the Semitic alphabet of the Phoenicians, but it developed a very different, oddly geometrical *ductus*. The Berber languages and dialects are Hamitic and are entirely distinct from Phoenician or any other Semitic languages such as Hebrew and Arabic.

It must not be supposed that because Tuareg women are bold Taureg men are not. I can give a good example of their reputation. One day when I was in an oasis village in the northern Sahara, word was passed from one end of the village to another: "A Tuareg is coming! A Tuareg is coming!" The streets emptied as if by magic, but I was probably not the only one who peeked. Marching down the middle of the street, veiled except for his eyes, looking neither right nor left, clad in a long black gown, came a majestic figure at least six feet tall — a great height in those parts. At his side was a tremendous straight sword, nearly as long as he was tall. He took what he wanted from a merchant's stall and strode back into the desert.

Another desert that I know and love is the desert, or rather the conglomeration of deserts, forming most of central Australia. I will come to that later when I write of some of our extensive travels in that enchanting continent. The only desert that has impressed me unfavorably is that of Atacama, along the coast of northern Chile. From the sea and again from the air it seemed to be completely bare of anything, including interest, but I have not set foot on it and if I did I might well find even it full of interest.

And then there is the sea. I did spend one year near the sea in California when I was a child, but except for that I grew up far inland and with little interest in the sea. On my first adult encounter, when I was twenty-one, it impressed me only as something quite unpleasant that had to be crossed to get to the other side. I traveled in steerage, or what, if it still existed now, might rate as about fourth class, and I was seasick almost all the time. That affliction continued through several Atlantic voyages and my first two visits to South America. It became airsickness on early plane travel. Then when I was in my thirties it suddenly stopped, and I have never been seasick or airsick since then.

When I was no longer seasick and when I could travel first class or on single-class ships, that became and still is my favorite means of travel. I enjoy just being on a ship and no longer view that only as a way to reach another land. How else can you be just as comfortable as at home and well waited on while traveling? As a boy I delighted in train travel, and my memory goes back to the days when Santa Fé trains stopped while the passengers dashed into a Harvey House and were served very rapidly by the famous Harvey Girls, doubtless charming but that was before I was fully appreciative of feminine charms. Now, like practically everyone else, I am disgusted with American railroads. Plane travel is the least comfortable of any, but it is also much the fastest and I use it constantly simply because otherwise I would not have sufficient time available on many occasions. And even when I can be more leisurely I fly to where I board a ship. Ships do not dock in Tucson, where even our great river, the Santa Cruz, is normally dry. And some ships I want to take do not sail from the United States. This becomes an increasing problem now that there are almost no regular passenger, as distinct from cruise, ship schedules.

And a Few Remarks on Cities

Kul blad u zinha ["Every country has its beauty"] — Moroccan saying published by Westermarck

Like Byron — although I suspect that he was not completely sincere — I have generally preferred mountains, deserts, and oceans to cities. I have always been based, so to speak, in cities small or large — Denver, New Haven, New York, Cambridge, Tucson — but have managed to spend about as much time away from them as in them. Now I cannot see any American city as beautiful or even very attractive in itself, not even San Francisco or New Orleans, which somehow keep that reputation. Paris charmed me when I was younger, but the charm has mostly worn off. It is a city for the young. London, not exactly beautiful, is endlessly interesting. It is a city for more mature appreciation. The English countryside is generally lush but has little to offer to a lover of mountains and deserts. I had often heard, for example, of the grandeur of the Cheddar Gorge, and kind English friends took me for a drive to see it. After we had ridden for some time I asked when we would come to the gorge and was told, more in pity than in disdain, that we had already been through it.

Some South American cities are exotic but I do not find any of them really endearing, and I must say the same of most African cities, although Cairo might be great if there were fewer Egyptians there. Among Australia's blessings are several charming cities: Canberra, Perth, Adelaide. In New Zealand the gardens of Christchurch make it outstanding.

Notes

§   One of my boyhood ways of making money was not exactly a business but may be sufficiently unusual to merit a note. During school
term on Saturdays, weather permitting, I used to take a streetcar on a line
ending near the South Platte River. There I panned the sand for gold. Later
I took the gold to the Denver mint, where it was solemnly weighed and
paid for. As I recall my take was almost always worth fifty cents and often
a bit more. My streetcar fare, round trip, was ten cents and I had to deduct
another ten cents as a payment on my gold pan, but that left what I considered a nice profit.

§   There is good precedent for calling Monte Rosa, highest in the Alps at
over 15,000 feet, the highest mountain in Europe. However, Elbruz in
the Caucasus is much higher, well over 18,000 feet, and is in Europe,
technically speaking. It is on the western side of the imaginary line that is
taken arbitrarily as dividing the Eurasian continent into Europe and Asia. I
haven't seen Elbruz, so I prefer to think of it as not *really* in Europe.

§   Speaking of Tuaregs, someone might like a short short story translated from the Tuareg (or from their language, often called
Tamachek):

A hound found a bone and was gnawing it. The bone said to him,
"I'm good and tough." The hound said to it, "Don't worry about it. I
have nothing else to do." [End of story.]

I am a fancier of Berber and Arabic stories, and I think that one is hilarious. Some in Arabic are even more simplemindedly funny.

§   Berber courting habits are illustrated by delightful Tuareg examples,
scratched on rocks relatively recently, for example:

"I am Timilla. I love the ardent lover who knows me well."
"I am Hennore. I want young men."
"I am Fatimatah. I want someone to take care of me."

— And a final cry from the heart:

"I am Fadis. I want someone."

§   Another peculiar thing about Berber writing is that tifinagh characters
can be written in any direction, right to left, left to right, up to down,
down to up, zigzag, boustrophedon, or even in a spiral.

§ My wife, Anne, has hardly seen one of my favorite cities, Christchurch, because by some strange fatality she has been seriously ill every time we were there.

References

Most of the things in this miscellaneous chapter are documented only in my memory and in various unpublished records. I do append a few basic references on the Sahara and its inhabitants, even though they are in French. The only others I know are in Arabic, presumably even less accessible to most readers.

Bissual, H. 1891. *Le Sahara Français*. Algiers: Adolphe Jourdan.

Hanoteau, A. 1896. *Essai de grammaire de la langue Tamachek'*. Algiers: Adolphe Jourdan.

Marcy, G. 1936. L'Epigraphie Berbère (Numidique et Sahariene). *Annales de l'Institut d'Etudes Orientales,* 2: 128–64. With a folding table of comparative alphabets.

12

Three Productive Years

It is work which gives flavor to life. — Amiel

I would not agree with that saying of the Swiss poet-philosopher if he meant that work is the only flavoring in life. A truly well-flavored life has many ingredients, but Amiel was surely right to include congenial and productive work among the most essential of them. Returning to chronological sequence, I now come to three years, more or less from mid-1939 to mid-1942, that were among my most hard-working and are seen in retrospect as also among the most productive.

In spite of the fallacy of the often quoted dictum "Publish or perish!" — which did not really obtain at the American Museum — publication is some indication of scientific activity, and my bibliography for 1940–42 has sixty-three entries. Some of these represent work done earlier, but some things written in these years did not become bibliographic items until later. That includes my first book-length work on evolution, perhaps the most influential one, and also a large work on classification in general with a detailed classification of mammals. Both were complete in manuscript when I entered the army in 1942, but the former was not published until 1944 and the latter not until 1945. Field work for the monograph on the Kamarakoto Indians, mentioned earlier, was done previously, but the monograph itself was written in New York in 1939 and published in Caracas in 1940.

Besides the research and exposition represented by those publications and in addition to other pressing museum work, it was also in those years that I was largely instrumental in the foundation of what has become a highly successful specialized professional society: the Society of Vertebrate Paleontology. With the exception of at least nine-tenths of the population, who have never heard the words or who have a vague idea that paleontologists have something to do with archaeo-logy, everyone knows what vertebrate paleontologists are and what they do. They are persons who study the ecology, associations, relationships, and histories of vertebrates, which are animals with backbones: fishes (in a broad sense), amphibians, reptiles, birds, and mammals. They do this primarily by discovering, collecting, preparing, and studying fossil vertebrates, but the study also frequently involves recent descendants and other relatives of the fossil animals. I have already given in outline an example of results of such work: the history of mammals in South America.

Like most scientists, vertebrate paleotologists concentrate their research on a part, only, of their general field. Thus my basic research has been mostly on mammals, although I have also made fewer but similarly intense studies of some reptiles and birds. In spite of individual specialization all vertebrate paleontologists share much in aims, attitudes, interests, methods, and needs. They are quite a cohesive group, not only in the United States, the country in which they are most numerous, but also throughout the world.

Sometime early in the present century — the exact date has eluded me — a small group united as the Society of American Vertebrate Paleontologists, acronymed in our hurried way as the SAVP. In 1907 that group discussed expansion to become a general paleontological society, bringing in specifically the invertebrate paleontologists (study-ing animals without backbones) and the paleobotanists (studying you-know-what). In 1908 a mixed group got together and organized such an expanded society, named simply the Paleontological Society (PS). Consequently the SAVP was formally disbanded and its members were incorporated in the PS, in which they formed at first about 35 percent of the membership.

The PS first met formally in 1909, has met almost every year since then, and is still going strong. But as the years passed the status and participation of the vertebrate paleontologists in the PS, which had been the outcome of their own initiative, steadily declined. In the country as a whole they were greatly outnumbered by other kinds of paleontolo-gists, especially the invertebrate paleontologists, who often had dif-ferent interests and who increasingly joined and dominated the PS. In 1933 a group among the most active vertebrate paleontologists undertook to consider what, if anything, might usefully be done about this. An informal poll was conducted by E. S. Riggs, who had been secretary of the SAVP when it disbanded. Those polled were overwhelmingly in favor of holding vertebrate sessions separate from those of the PS, but were almost evenly divided on continuing as a part of the PS or forming a separate society. It was decided to remain in the PS but to organize as a separate section.

Such a section, self-constituted, was set up in 1934 and was even recognized as such by the PS, which published proceedings of the Vertebrate Section, but in fact there was no provision for such sections under the PS constitution. The PS adopted a new constitution in 1937 which did provide for the establishment of sections by petition of twenty or more fellows and members of the PS, but the vertebrate paleontologists never petitioned under that provision. One reason for that was that a number of vertebrate paleontologists active in the section were not and did not wish to be members of the PS, a requisite for formal sectional membership. Another reason was a growing feeling that a separate society would eventually be desirable or indeed necessary.

The Section on Vertebrate Paleontology was active from 1934 to 1940. It had two officers, a chairman, who held a more or less honorary position with no duties except to preside at meetings, and a secretary, who did whatever active work was needed. A. S. Romer ("Al") was secretary from 1934 to 1938, and I was secretary in the transition years 1939 to 1940. We met once a year, sometimes in rooms adjacent to those of the PS but more often at a different place in the same city or in a different city, the selection depending on whether there were vertebrate paleontologists and collections of fossil vertebrates in the cities involved. The PS always met, and still does meet, with the larger and more inclusive Geological Society of America (GSA), which (quite properly) does not have local interest in vertebrate paleontology among its criteria for meeting places.

The roster of chairmen of the section should be on record. In sequence it is: Chester Stock, Elmer Riggs, Remington Kellogg, Walter Granger, Barnum Brown, Le Roy Kay, and Al Romer — all distinguished vertebrate paleontologists, all older than most of us, but good friends of ours, and all now dead.

In 1939 and 1940 it became obvious that the anomalous situation of the section could not continue. The vertebrate paleontologists would have to organize formally as a section of the PS under the requirements of its constitution or set up formally as a separate society. As secretary I circularized previous participants in the section and others possibly interested. After considerable correspondence and work on my part it became clear that there was now almost unanimous support for transforming the section into a Society of Vertebrate Paleontology, friendly to both PS and GSA but not formally affiliated with them. I therefore drew up a tentative constitution for such a society and assembled data on its needs and outlook.

On 28 December 1940 the nominal Vertebrate Section of the PS met for the last time in the Museum of Comparative Zoology in Cambridge, Massachusetts, under the chairmanship of Al Romer. I presented the plan for the proposed new society and it was unanimously adopted, with the proviso that the organization be provisional, only, until a meeting in the following year, so that many vertebrate paleontologists not present at the organizing meeting could be charter members and have an opportunity to attend the first regular annual meeting of the new society. Until that meeting, Al Romer, as the current section chairman, would keep on as provisional president and I, as the current section secretary, as provisional secretary-treasurer. The first regular officers were elected by mail vote to take office on 1 April 1942, after the first annual meeting, which was held on 29-30 December 1941 in Boston and Cambridge as arranged by the provisional officers.

The SVP began as a small organization. At the time of the first annual meeting it had 229 members, all but one of them charter members. Early meetings were usually attended by thirty or forty members, practically all well known to each other. Meetings then were delightfully informal and the scientific sessions consisted not of reading technical papers but of calling on each person present to get up and tell us what he or she was doing. Often one just happened to have a particularly nice specimen in a pocket, calculated to make the others jealous. The social sessions were boisterous but only occasionally rowdy. The success of the society was the death of that delightful informality, a loss often and loudly bewailed by Al Romer in particular. The membership and the meetings became far too large for such disorder, and formal papers are now read and even social occasions arranged on a tight schedule.

Although I have omitted much in this account, I may have given the SVP more space than will seem called for by readers who are not members of that organization. There are two fairly good reasons for that. First, the society has been of great importance to me and this is, after all, an autobiography. Second, I want to set the record straight on some particulars. Published accounts, including a history of the GSA and its affiliates as well as some unchecked memories of an early officer of the SVP published in its own bulletin, have not been entirely accurate.

So that's that. I will now change the subject, and although the chapter epigraph is still relevant I will add a new one.

I have no great quickness of apprehension or wit which is so remarkable in some clever men, for instance Huxley. — Charles Darwin in his *Autobiography*

That is one of three things that I share with Darwin, although if applied to me the Huxley would be the grandson of Darwin's Huxley. My second Darwinian trait is devoted and long-continued study of the phenomena and principles of organic evolution. The third is that I have been ill during much of my life. It would be pointless and dull here to relate details about my first and third Darwinian traits. I now turn to the second, which has already been exemplified and will be again.

This has involved great interest in Darwin himself. I have read all his books and followed *The Origin* through its several editions. I have also read almost all the published letters, biographies, and other studies bearing on Darwin's life and work. I have made what could be called pilgrimages to the room where Darwin was born and the room where he began his youthful explorations into natural history, both in Mount House, Shrewsbury; Shrewsbury School, where he received most of his primary education; Christ's College, Cambridge, where he completed his higher education; Down House, Downe, where he spent most of his adult years, wrote *The Origin* and much else, and died; his tomb in Westminster Abbey.

On my living room wall there is a color photograph of an unpublished portrait of Darwin painted by Laura Russell on 23 August 1869. Laura Russell, wife of Lord Arthur Russell, was visiting Lord and Lady Derby, who were friends of the Darwins and lived not far away. The Russells and Derbys visited Darwin at Down House and the portrait was painted there. The photograph was given to me by Lady Nora Barlow, granddaughter of Charles Darwin, when Anne and I were visiting her in 1965. It is a focal point at a celebration of Darwin's birthday with a group of graduate students, which has become an annual event. (This is to us a greater event than the birth of Abraham Lincoln, which also occurred on 12 February 1809.)

The Origin of Species accomplished two great things. First, it supplied so much evidence and such calmly convincing reasoning that soon after its publication the great majority of biologists accepted evolution as being, in a colloquial sense, a fact. Second, it went far toward providing a reasonable and testable explanation of that "mystery of mysteries," the origin of species, and indeed not only of species but of all the groups of organisms with their intricate interrelationships. The only rational argument for the divine creation of species has been the evidence that all organisms are adapted for life where and as they live it and that it is extremely improbable that such adaptation could have arisen by chance alone. Darwin agreed, and he supplied the essentials of an explanation, of a nonrandom, antichance factor that involves no divine or other arcane intervention: natural selection.

It is well over a century since *The Origin* was published and knowledge of biology has since vastly increased. The reality of evolution is even more firmly established than in Darwin's day, but inevitably and desirably evolutionary theory or causal explanation has become more complex, more complete, and in some respects different from Darwin's own views. Darwin tentatively advanced a hypothesis about heredity that is now known to be incorrect. Darwin recognized that possibility but correctly saw that any hypothesis based on facts of heredity already known to him could supplement but not supplant the explanatory force of natural selection. It has been said that Darwin's great weakness was failure to take Mendel's work on heredity into his theory. That seems to me quite incorrect. Mendel's work was not published until years after *The Origin* had appeared and after it had been twice revised. The legend that Mendel's publication was so obscure that it remained virtually unknown is not true. It was soon known to a number of prominent biologists and may have been known to Darwin, although there is no positive evidence to that effect. The point is that neither Mendel himself nor anyone else at the time saw that Mendel's discoveries had any bearing on evolution.

The somewhat fallaciously so-called rediscovery of the Mendelian principles in 1900 had a curious effect on the study of evolution. Geneticists did then see that Mendelism and other discoveries in their field had a bearing on evolution, but some, for a time an apparent majority, concluded that evolution resulted from random mutations. Natural selection, they thought, was only a negative factor that might eliminate grossly inadaptive mutations but could not cause, intensify, or change adaptation. In effect that was a return to the view that seemed incomprehensible to Darwin and has always seemed so to most naturalists: that adaptation arises at random.

When I was in graduate school and in my early years at the American Museum my teachers and colleagues all accepted the truth of evolution but few of them were Darwinians, or neo-Darwinians, in any precise sense. Most of those who were theoretically-minded at all were divided between the geneticists' view of evolution by random mutation and what was incorrectly called neo-Lamarckism, evolution by the inheritance of acquired characters. My major professor, Lull, taught a highly popular course on evolution and wrote what was then the leading textbook on the subject. In effect he gave equal billing to all the conflicting theories on the causes of evolution (there were several others besides the main three that I have mentioned), but he personally espoused none of them.

Despite all that confusion, there were biologists who continued to think and work on Darwinian, that is, selectionist grounds. A strengthening and widening of that position began in the 1930s and within the general scope of the Darwinian revolution. It led to what is now commonly called the synthetic theory, because it became a synthesis from all the many branches of biology, including paleontology. Having so many sources, it cannot be ascribed to one or even to a few students, but its beginning was apparent in works published by R. A. Fisher in 1930, by Sewall Wright in 1931, and by J. B. S. Haldane in 1932.

I have mentioned that increasing concern with evolutionary theory was one of the doors opening to me in the 1930s. Most paleontologists had been little concerned with theoretical or in any broad sense interpretive matters. With rare exceptions invertebrate paleontologists tended to scorn such studies and to cling to descriptive systematics and stratigraphy. Among vertebrate paleontologists E. D. Cope, who died before I was born, was a thoughtful interpreter and theorist but was a neo-Lamarckian, an attitude still respectable in his day but fully discredited in mine. H. F. Osborn, whom I knew well in his old age, invented an idiosyncratic theory of evolution tending (under influences that he left extremely vague) constantly toward better, more aristocratic ends. He was resigned to the fact that no one accepted his views during his lifetime, and they have simply been forgotton since then. W. B. Scott, whom I also knew well in his old age, had written some interesting theoretical studies in his thirties, which were in the 1890s, but thereafter he decided that evolution, although a fact, was simply inexplicable on Darwinian or any other grounds. Louis Dollo, whom I barely knew personally although I was a practicing paleontologist and had spent considerable time in Europe before he died in 1931, was probably the most thoughtful paleontologist in Europe. His contributions were, however, mainly in the form of apothegms which described or defined so-called laws rather than attempting to explain.

Thus, as I saw a possible synthesis beginning to take form, I also saw a serious gap in it. Paleontology as the study of the history of life should provide if not *the* surely *a* touchstone for the nature and validity of evolutionary theory. My zeal was awakened not only by the already mentioned works of Fisher, a biometrist, Wright, a population geneticist, and Haldane, catalogued as a geneticist but more a sort of universal biologist. I was even more stimulated and owe most to Dobzhansky's book *Genetics and the Origin of Species*, which appeared in 1937. That great, seminal work showed me that genetics is indeed consistent with and partially explanatory of nonrandom explanations of evolution. In 1938 I began work on a book that would relate paleontology to this point of view. I was also stimulated to a lesser degree and in quite the opposite way by a work by Schindewolf, a German invertebrate paleontologist, who had first sought a synthesis between paleontology and genetics in 1936 but who adopted the to me entirely unacceptable view that evolution is essentially random and involves mainly mutations effecting changes early in ontogeny, individual development. Much the same might be said of a book by Goldschmidt, an eminent geneticist, who in 1940 carried to an extreme the already increasingly dubious view that evolution is mainly a matter of radical, random mutational modifications of the whole genetic system.

My book, *Tempo and Mode in Evolution*, was completed in 1942 and was published without opportunity for further revision just after I returned from army service overseas in 1944. There is little point in an autobiography's repeating or summarizing the substance of a book widely available, but I will quote my own summary made recently when asked my own opinion of some of my work:

"I am keenly aware that much of *Tempo and Mode in Evolution* now seems primitive and that parts of it have been invalidated. Nevertheless, I am consoled by the conviction that it had some historic value, that it was a success at least to the extent that it did bring in a new field of study and a new thesis into the development of the synthetic theory, and that its thesis has stood up well. That thesis, in briefest form, is that the history of life, as indicated by the available fossil record, is consistent with the evolutionary processes of genetic mutation and variation, guided toward adaptation of populations by natural selection, and furthermore that this approach can substantially enhance evolutionary theory, especially in such matters as rates of evolution, modes of adaptation, and histories of taxa, particularly at superspecific levels."

That book was one of a remarkable series of contributions to the synthetic theory of evolution from Columbia University Press that may be said to have established that theory on a firm and truly synthetic basis. It began in 1937 with Dobzhansky's book already mentioned, continued in Mayr's *Systematics and the Origin of Species* in 1942 (published before my book in the series, but not seen by me until mine was also published), then mine in 1944, and finally in 1950 Stebbins's *Variation and Evolution in Plants*. The same press brought out revisions of Dobzhansky's book in 1941 and 1951 and then in 1970 a revision so complete that it was retitled *Genetics and the Evolutionary Process*. Mayr also wrote what could be considered a completely rewritten and retitled version or successor of his book, published in 1970 as *Animal Species and Evolution* by Harvard University Press. I also reworked *Tempo and Mode* so completely that the revision was retitled *Major Features of Evolution* when published by Columbia University Press in 1949.

The other big work written in this period is *The Principles of Classification and a Classification of Mammals*, completed in 1942 and published by the American Museum in 1945. It occupies 365 large octavo pages, and includes an essay on taxonomy (33 pages), a then new classification of mammals, living and extinct, down to genera (129 pages), historical and other notes on the classification (110 pages), bibliography (30 pages), and index (42 pages). I am grateful to say that colleagues at the American Museum checked references, made the index, and prepared the final typescript while I was in the army, but I did all the rest single-handed. For many years this was the standard reference work on its subject, and it still is today *faute d'autre*. Since it was published a tremendous amount has been learned about mammals, especially fossil mammals, and my classification is far out of date. I had intended to revise it, but when I left the American Museum in 1959 I necessarily abandoned that plan. The staff of that museum has ever since then been working on a new classification, not properly speaking a revision of mine, but this is not near publication as I now write (late 1976). I have misgivings not about the quality of the work but about the taxonomic principles being followed, which I consider inappropriate for this particular use.

Before closing this rather sober chapter I should just mention a few other things that I did and published in this period. They include eight short studies of Paleocene and Eocene primates, the oldest of our relatives then known (a few somewhat older have been found since). By the way, on the strength of that and some other studies, including the monograph on the Kamarakotos, I applied for fellowship in an anthropological association, but a colleague in the Department of Anthropology at the American Museum blackballed me because I wasn't in his department and therefore not a *real* anthropologist. I was still working on South American faunas and faunal history, of course, and made some contributions to straightening out and modernizing the stratigraphic nomenclature for the mammalian succession in Argentina. For a study of some American fossil cats I devised a graphic method of comparing skeletal proportions by ratio diagrams, now widely used for all sorts of animals.

The study of fossil cats was one of several outgrowths from the discovery of some jaguar remains in a cave near Sweetwater, Tennessee. The cave belonged to two gentlemen in that town who brought the bones to New York for identification. The bones were so interesting that in May 1940, Anne and I went to Sweetwater to examine the occurrence, look for more bones, and make a plaster cast of footprints left in soft mud by the poor jaguar that had gone astray in the dark depths of the cave and never got out. Cave crawling, or spelunking, does not really attract me, but bones do and I would go anywhere, even in the depths of a cave, for a good one. An agile young man familiar with the cave crawled with me, at one place in a crevice so small that my pants were peeled off me as I squeezed through. On another crawl one of the owners started with us, but when things got really hairy he remembered an errand elsewhere, backed out as fast as he could, and let us go without him. When I twitted him a bit about that afterward he said, "Well, I wouldn't have minded so much if I hadn't suddenly remembered that the boy is an epileptic and we'd 'a' been in a right bad fix if he had a fit!"

I also made an extensive study of some sabertooth cats and how they used their sabers. A Swedish paleontologist disagreed and we had a friendly controversy for some time. I still think I was right! I also proved to my own satisfaction that continental drift, if it occurred at all, had no effect on the distribution of Cenozoic mammals. I was wrong about that, but not as wrong as some of the drifters, whose ideas were pretty primitive at that time.

Also in this period (do not despair; I am not going to discuss any more of the sixty-three bibliographic entries) I had much pleasure in researching and writing a history of early vertebrate paleontology in

North America. This was published by the American Philosophical
Society, which gave it a prize graciously accepted for me by my wife
while I was overseas during Hitler's war, as will be briefly related in the
next chapter.

Notes

§ Paleontologists and archaeologists are usually confused by nonscientists, to the annoyance of both. In origin the words *paleontology* and *archaeology* could mean the same thing: Greek *palaios,*
"ancient," *onta*, "existing things," *-logia*, "discourse or study"; and
archaios, "ancient," plus *-logia* again. Nevertheless these coined words,
not used by the ancient Greeks themselves, have always had quite different meanings in modern languages. *Paleontology*, probably first used
by Lyell in 1838, has always meant not the study of anything ancient
but the study of any traces of ancient living things, that is, of fossils.
Archaeology, an older term already in use in the seventeenth century,
originally referred rather vaguely to almost any study of ancient human
history but has now long been confined to the study of old human artifacts: tools, buildings, and the like. Students of very ancient human
remains, especially before *Homo sapiens*, are paleontologists of a very
specialized sort, but are a minute minority among paleontologists in
general. Students of *Homo sapiens* as organisms are physical
anthropologists.

§ Membership requirements for the SVP are simply that members be
adult and interested in the subject. Members of the PS are expected
to have paleontology as a part, at least, of their professions. The PS
considers all of paleontology, including vertebrate paleontology, to be
within its scope. A number of vertebrate paleontologists belong both to
it and to the SVP, but rarely take part in PS meetings. The PS publishes
numbers of the *Journal of Paleontology* alternately with the Society of
Economic Paleontologists and Mineralogists and sometimes includes
research papers on vertebrate paleontology. The SVP publishes a news
bulletin, noted below, which does not include research papers.

§ In Fairchild's history of the GSA, cited below, the PS is designated
as a branch, offshoot, and child of the GSA. It is true that the PS
became affiliated with the GSA soon after its organization, was for a
time considered a section of the GSA, and continues close association

with the latter. Nevertheless the original impulse for formation of the PS came from the vertebrate paleontologists and if the PS is the off-spring of anything it is that of the prior vertebrate organization, the SAVP. It even inherited the (extremely slender) treasury of the SAVP. As the SVP was originated by a section of the PS, although the section was informal, illegal, or at least unconstitutional, it could be considered an offshoot of the PS, but the latter still lives and the SVP has inherited nothing from it. I prefer to think of it as a resurrection of the SAVP, now with many foreign members and not strictly American, although it meets only in North America and has only North American officers.

§ I met Dobzhansky, then at Columbia University, at about this time, and we became close friends and so remained until his death in 1975. I was also associated with Mayr at the American Museum and later at the Museum of Comparative Zoology. And I became well acquainted with Wright and Stebbins among the early syntheticists whom I have named in this chapter, although I was not so closely associated with them as with Dobzhansky and Mayr. I also met Schindewolf and Goldschmidt, antisyntheticists, briefly. I never met Fisher, who seems not to have been very cordial toward younger and less eminent colleagues (He was twelve years older than I).

References

Barlow, N., ed. and annotator. 1958. *The autobiography of Charles Darwin, 1809-1882*. London: Collins. This autobiography, not intended for publication, was published by the family with numerous omissions shortly after Darwin's death. The 1958 edition, edited and with extensive notes by Darwin's granddaughter, was the first publication without omissions. An identical American edition was published in 1959 by Harcourt Brace Jovanovich, New York.

Darwin F., ed. 1887. *Life and letters of Charles Darwin*. London: Murray. The editor was Charles Darwin's son. Many more letters have since been published, but these cover the subject adquately for most readers' purposes. A more accessible edition, also in two volumes, was issued by Basic Books, New York, in 1959, with a foreword by me.

de Beer, G. 1964. *Charles Darwin*. Garden City: Doubleday. Fairly brief, but perhaps the best recent biography.

Dobzhansky, Th. 1970. *Genetics of the evolutionary process*. New York and London: Columbia University Press. The bibliography, pp. 433-84, contains references to practically all the important literature of the synthetic theory in its formative years.

Fairchild, H. L. 1932. *The Geological Society of America, 1888-1930*. New York: published by the society.

Huxley, J., and H. B. D. Kettlewell. 1965. *Charles Darwin and his world*. London: Thames and Hudson. A sketchy but authoritative biography accompanied by a large number of illustrations.

Peckham, M. 1959. *"The origin of species" by Charles Darwin: A variorum text*. Philadelphia: University of Pennsylvania Press. Ordinary editions of *The Origin* are so numerous and so widely known and available that I do not bother to cite one here. They are mostly reprints of the original sixth edition, the last one revised by Darwin, although there are at least two reprints of the first edition, one, reset, by the Philosophical Library (New York, 1951), and one in facsimile by the Harvard University Press (Cambridge, Mass., 1964). I cite Peckham's edition because it is the only variorum, clearly showing the successive changes from the first edition made in all later editions and thus indispensable for any serious student of Darwin's work and thoughts.

Simpson, G. G. 1942. The beginnings of vertebrate paleontology in North America. *Proceedings of the American Philosophical Society* 86: pp. 130-88. This covers the period up to 1842.

————. 1944. *Tempo and mode in evolution*. New York: Columbia University Press.

————. 1945. The principles of classification and a classification of mammals. *Bulletin of the American Museum of Natural History*. 85 (whole volume).

Society of Vertebrate Paleontology. 1941- . *News bulletin*. The history of the society, lists of its members, and the activities of many of them are recorded in the *News Bulletin*. The first seven issues were mimeographed but following issues are reproduced by photolithography from typed copy. A fixed schedule has not always been followed, but there are now three issues per year. The October 1976 issue was no. 108.

13
1942–1944

À la guerre comme à la guerre. — attributed by Dorothy Sayers to her character Lord Peter Wimsey as a French saying in the First (the Kaiser's) World War. Perhaps she made it up herself. It is apt, but cannot be literally translated. "When at war one acts as if at war" is a fair translation of the sense.

I had an impulse to head this chapter "1942–1944" as I have, then to leave a page blank and go on to other dates and things, but that would be cheating. In this period I spent two years as a captain and major of the Army of the United States, almost all that time in what is sardonically known as the Mediterranean theater — there were few comic skits in this theater, although I will relate two. To be fair I must somehow account for these years of my life.

By reference to the file of orders and assignments kept by any officer, I find that Army Serial Number 0920668 had fifteen different assignments to duty from 3 December 1942 to 7 October 1944. The first was training — all of two weeks! Four more brief assignments had to do with staging, awaiting orders, processing, and terminal leave. Two assignments were incidental to movement from the United States to North Africa as cargo security officer on a Liberty ship and return from North Africa to the United States on an army transport, in command of troops whom I had never seen before and never did again.

The other eight all concerned intelligence and counterintelligence, generally with Allied forces including British personnel, and for a time as a chief liaison officer with free French personnel in North Africa. I moved from Algeria to Tunisia to Sicily to mainland Italy and back to Algeria. None of that was cloak and dagger, all in uniform. It is fairly typical of the army that there were special insignia (or a special insigne) for military intelligence but officers actually engaged in military intelligence were not allowed to wear them and were obliged to wear infantry insignia instead.

My duties were terrifying at times, boring at others. I do not intend to relate what they were precisely or much of what happened to me. That is through no consideration of national security '— my orders rarely had a classification higher than Confidential, which amounts to no classification at all. On looking over them again, without nostalgia, I do see "SECRET (Equals British MOST SECRET)" or what my British colleagues called "TOO, TOO SECRET," but that did not happen often. My activities did no apparent harm to the war effort and I think that they helped, but they are not of historical interest. As for my biography, they are also quite outside the mainstream of my life. It is a psychological factor that they were on the whole unpleasant and that it is a human, forgivable, even perhaps commendable trait not to dwell unduly on past unpleasantness.

I will therefore now relate two incidents, neither with any direct connection with my army activities, one because it is funny and mysterious, the other because it is funny and not mysterious and received some unwanted publicity at the time.

One day in Algiers, in connection with liaison duties, I had occasion to visit a French officer in the hills outside the city. He suggested that since our business might take some time I send my jeep and driver back to the pool and he would provide transportation to my billet. One of the innumerable silly regulations was that commissioned officers were not allowed to drive cars. In due course I embarked in the French officer's jeep with his driver, one of the picturesque Berber soldiers known as *goums* in the Free French army. Of course he had no English, and this particular goum had no French either, but he did speak some Arabic. All went well until we were in the city driving down the main street. Then pandemonium broke loose. Gendarmes blew whistles and waved truncheons at us. MP's drew weapons as they came running after us. Wood-stoked buses pulled up and robed civilian Arab riders jumped out and sought shelter. I told the driver in my fractured Arabic to slow down, let me jump out, and then drive like hell (which is *jahannam* in Arabic, cognate with our *gehenna*). He did; I did. I rushed around a corner and then strolled nonchalantly and unchallenged into one of the lower and safer, though nevertheless out of bounds, alleys of the casbah. That's all. I never found out what the uproar was about.

The publicized episode occurred while I was attached to General Patton's headquarters in Sicily for mess and billet. I was wearing a beard, as I had sporadically before 1935 and have continuously since then. Like many men perhaps worried about their own machismo the violent general detested beards. He sent an aide, a chicken colonel, to tell me that if I did not shave my beard he (Patton) would pluck it out hair by hair. With perfect military courtesy I explained to the colonel that according to army regulations a commissioned officer could wear a neat beard if he had permission from his commanding officer, and that General Patton was not my commanding officer. General Eisenhower was, and I had his permission to wear a beard. If General Patton desired the removal of my beard he would have to take the matter up with General Eisenhower, who was also his commanding officer. Unfortunately the colonel had not bothered to get me alone when delivering Patton's ultimatum, so a correspondent got hold of the story and sent it back to the United States, where it made quite a stir — Patton was widely and understandably disliked by correspondents. Not long thereafter they had an even better crack at Patton when he slapped a sick soldier (I happened to be in the same hospital at the time) and was ordered by Eisenhower to make a public apology.

I received no apology, but I did keep my beard. On Patton's orders I was thereafter followed everywhere by MP's trying to catch me out. One did once find me technically "out of uniform" (the edge of a sweater knitted by my wife showed at the neck of my blouse), but the several who thought they had found me unarmed in a combat zone were fooled: I wore my pistol in a shoulder holster and not in the more usual way on my belt.

Some people have made a hero out of Patton, but I think the best thing he ever did for his country was to get killed (not by enemy action) without returning to the United States.

Thinking it over, I realize that the impact of war cannot be passed over with mere mention of two absurd incidents. War has thrilling moments and it has moments of terror and of disgust. I want some minimal exemplification of these more serious aspects, and I have difficulty in making a choice.

An incident in Tunis will do as an example of a thrilling event. I had entered Tunis quite early, in fact inadvertently before the Germans had completed evacuating it. When the Germans were indeed gone, there was a triumphant parade not of our troops but of Leclerc's French goums. Against all odds these men had fought their way right across the Sahara and those who survived were ending the extraordinary march with entry into Tunis. Seeing these hard cases, barefoot or sandaled, in long robes, swinging by to the music of shrill flageolets was a moving sight that still brings me almost to tears when I think of it.

Incidents of fear were numerous, and I will go no farther than my first real duty assignment for an example. I have mentioned that I went overseas as cargo security officer on a Liberty ship. Each of the officers in my unit had been put on a different ship, in anticipation that some, at least, would make it. Those ships had civilian crews who were not allowed to see them loaded. Only the cargo security officer had a sealed inventory of the cargo. My deck load was obvious enough: it included two locomotives, and the first mate and I spent a horrible night on deck in a heavy storm prepared to cast them overboard if they pulled out the deck plates to which they were chained. But even I had no knowledge of the hold cargo, as my orders were not to open the inventory except in a serious emergency.

Our convoy was scattered in the storm and lost its escorts. Our ship, its engines in poor shape, was limping slowly along on its own as one by one other ships that had been in the convoy broke radio silence to announce that they had been torpedoed by U-boats and were sinking. I decided that this was a legitimate emergency and opened the sealed inventory. I then told the captain not to worry about what to do in case we were torpedoed. Our holds were full of aviation gasoline and high explosives.

Incidents of disgust were not as numerous as those of fear, but they do offer some choice. At one point I heard that things were amiss with a unit that had gone ahead of me because I was disabled. I was carried to a plane by two husky GI's and flown (as usual in an unarmed plane within range of German fighter fields) to union with the unit. Discrete investigation convinced me that the trouble arose from a handsome but perfidious woman, and she was eliminated. By nature I am among the most kind-hearted of men, but "à la guerre comme à la guerre."

I finally acquired enough points to be rotated home, and on arrival in the United States I was ordered to inactive duty. Within a few days an officer with whom I had worked in the Mediterranean telephoned me from the Pacific and told me that if I would return to duty there he could promise my immediate promotion to colonel. I told him that I would of course obey an order to do so, but that I had no ambition to become a colonel and would not volunteer. He dropped the matter. Some time later I was informed that I would be officially discharged if I joined the reserves, but I turned that down too. In fact I never did receive a discharge and as far as any official record goes I am still an officer of the Army of the United States on inactive status — unpaid and unpensioned of course.

I still feel that our war against the Germans, Italians, and Japanese was both just and necessary, and I have no ethical qualms about having volunteered and served in it. Nevertheless it left me with a permanent and profound hatred of warfare and with an equally permanent and profound distrust of professional soldiers, ours or any others — the higher in rank, the more the distrust.

I thought from the start that our war in Vietnam was neither just nor necessary, and I condemned it in speech and in writing long before that became the general opinion. I love my country, and it was heartbreaking to be ashamed of it.

Notes

§ I see by my authorization for a campaign ribbon that the chairborne soldiers back in Washington referred to the "European African Middle Eastern Theater," but we called it the Mediterranean theater.

§ Probably the reader has thought of the most likely explanation of the mysterious episode in Algiers: that the French officer had stolen the jeep and sent me down to town in it to check whether it was being sought. I thought of that one and found that it was not true.

§ When I joined the army my wife, Anne, who likes beards in general and mine in particular, looked up the army regulation concerning them. Thus when the question arose I could immediately cite chapter and verse to Patton's aide.

§ As a hardened etymophile I have tried to trace the word *goum* to its source, impeded by inadequate knowledge of Hamitic languages. *Goum* is the French form; in English "goom" would more nearly represent the pronunciation. French dictionaries indicate that the word is of Arabic origin, but that is improbable because the sound of hard *g* does not exist in the Arabic dialects spoken by the goums themselves. (It does exist in the Egyptian dialect, but no goums were Egyptian.) Dictionaries of Algerian and Moroccan Arabic do have the word *goum,* written with a special character for hard *g* used only in a few words of non-Arabic orign, but they do not give its origin. Hard *g* does frequently occur in Hamitic words and there is a letter for it in the old Tifinagh alphabet. Kabyles, the only Hamitic speakers with whom I have had some familiarity and incidentally the ones most familiar to the French, have a plural noun *aguam* meaning "people." The singular is pronounced "goom" (modern Kabyles have no alphabet). They do apply that to a native soldier, and I believe that is the source of *goum*. The Kabyles also have a more specific word *aa'sekriou* for "soldier," but that is obviously their mispronunciation of the proper Arabic word *askar*. (During later travels in black Africa I found that the same Arabic word has been taken over into Swahili and converted into *askali*.)

§   The goums were splendid men to have on your side, horrors to have against you. One of their favorite amusements was to slip into the German lines at night, find three Germans sleeping next to each other, and cut the throat of just the middle one.

§   A member of the crew on my Liberty ship was in bad shape before the voyage ended. Although blissfully ignorant of the nature of our cargo, he realistically assumed that we might all be killed at any moment, and he did as well as he could to make sure that he would not go to heaven with his pants down.

14

My Career as a Professor,
or The Universe and Other Things

The halcyon period between the self-tormenting exuberance of youth and the fretful carpe diem *of approaching senility.* — Dorothy Sayers

While I was a professor at Columbia University I had an office there in Schermerhorn Hall, an edifice noted, on one hand, for the fact that *Drosophila* genetics started there and, on the other, for the fact that the Manhattan Project, which eventually resulted in the atom bomb, also started there. My office had no door to a corridor but was approached through a couple of other rooms. Few people knew that it existed and almost no one knew that it had an occupant. That made it quite unsuitable for consulting colleagues or cajoling, advising, and admonishing students, but it was almost ideal in another way: no one ever interrupted me there. That did have one drawback: one night a watchman, unaware of my existence, locked the far corridor door when I was the only person left in the building. There was a telephone, but it took a long time to wake an operator and yet longer to pursuade him that this was not a practical joke and that there was indeed a bona fide professor immured in the fastnesses of Schermerhorn.

Much of my time in that hideout was devoted to writing a book about horses, titled, with an elegant concision not always characteristic of me, *Horses*. (I later wrote a book with an even shorter title, *Life*, as I will relate in due course.) Long years before, W. D. Matthew had written for popular sale at the American Museum a pamphlet on the evolution of the horse family. This had become out of date, and I was asked to revise and rewrite it. I became all too interested in the task. I already had a fair grasp of the literature on fossil horses, but I decided that I would also include living horses, about which my knowledge was limited.

If this were a chapter on my career as a horseman it would have to go back to 1924 when I first did much riding. That summer Charley Falkenbach and I hired horses from Santa Clara Pueblo in New Mexico in order to explore badlands in the Santa Fe formation where our model T Ford could not be persuaded to operate. My horse somehow sensed that I was neither an Indian nor a horseman and that there was a simple way to rid himself of the burden. Once riding down the main street of Espanola, nearest town to Santa Clara Pueblo, hardly a metropolis but with an interested audience, I zigged when the horse zagged. I turned a complete somersault, somehow came up standing on my feet with a cigarette still in my mouth (I was a heavy smoker in those days), and had enough aplomb to blow a smoke ring to loud applause. The horse's disapproval of me did once really do me a good turn. Out in the badlands the beast decided that if I wanted to go down a steep slope I could do it without him, which I did head first and landed beside the best fossil we found all summer — the perfect skull and jaws and most of the skeleton of the bearlike dog *Hemicyon ursinus.*

This is not, however, a chapter on my equestrian feats, so I return to professorial and curatorial matters. In pursuit of knowledge about living horses I not only read widely but also corresponded with all the American breed associations. I finally had so much material that it literally filled a book, and I am afraid that the pamphlet never did get revised. The book was published in 1951 by Oxford University Press (New York, not England). The book had a modest success, about what novelists tend to pass over as a *succès d'estime,* although it was later reprinted in both hardback and paperback. In one respect it has been something of an embarrassment to me, in somewhat the same way as Scott's medal for exploring Patagonia was to him. Some people, feeling called upon to say something nice about me, have referred to my brilliant researches on horses. The truth is, as I thought was clear even in the book, that I personally have done little research on horses and that little was not extraordinary. Almost all the information in the book came from others, either in printed form or in correspondence.

Once more back to my career as a professor. I became one in 1945 when I was appointed a professor of zoology at Columbia University and I have been a professor, of sorts, ever since. In fact for a time I was three professors, then two as the sum of one and a half plus a half, and as I write this I am a whole professor emeritus and half an active professor. When placed on inactive status by the army, I rejoined the American Museum staff as curator of fossil mammals and birds. Most of the time I was also chairman of the Department of Geology and Paleontology and from time to time also dean of the scientific staff, a rotating elective position. Columbia paid the museum, not me, for my services, and the museum simply paid me a salary as a curator.

In the years 1945 to 1959, besides other professional duties I nine times gave a graduate course relating to mammalian evolution. The classes were held at the museum in the fifth-floor room where Will Gregory had also taught Columbia classes for many years. It was a disadvantage that unless the students were working at the museum, which was true only of the few who were taking (or hoping to take) degrees in vertebrate paleontology, I saw nothing of them outside the formal classes. I left both the American Museum and Columbia in 1959 and went to the Museum of Comparative Zoology (M.C.Z.) and Harvard University, about which more later. From 1967 to 1970 I was connected both with the M.C.Z. and the University of Arizona, and since 1970 I have been a half-time professor at the latter institution only.

For me the time from 1944 to 1956 falls into what was so neatly characterized as the "halcyon period" by Miss Sayers, whose later and more serious work does not much attract me but whose Lord Peter novels do. These years were also ones of greatly varied activity for me, so much so that I am again reminded of the professor who was wont to lecture on "The Universe and Other Things," which I have borrowed as an appropriate supplementary title for this chapter.

So on with the other things, first the foundation of another society. I was active among the founders and was the first president of this society, as I had been of the Society of Vertebrate Paleontology. Nevertheless I shall discuss this event more briefly because I was not the leading promoter of the Society for the Study of Evolution, as I was of the SVP, and in this case I do not know of any misstatements that need to be set right. Ernst Mayr was the effective leader in the group that initially (before the war) was the informal Society for the Study of Speciation but that broadened its scope and changed its name to include all aspects of (organic) evolution when it was formally founded in March 1946. The society has held annual meetings ever since, but its main activity is the publication of a quarterly journal, *Evolution: International Journal of Organic Evolution.* This was started in 1947 with an enabling grant from the American Philosophical Society — I did arrange that. Mayr was the first editor and continued in that onerous position for some years. There has always been a board of eighteen associate editors, six elected each year to serve for three years. On the first board no

fewer than ten associate editors were Europeans and both those editor-
ships and the membership have always been international, although for
practical reasons the officers are American and publication is in English
only. The society has been highly successful and its journal, now (1976)
in its thirtieth volume, has flourished and early became self-sustaining
through society dues and subscriptions.

Instrumental in the formation of the Society for the Study of Evolution was a group cumbrously called the Committee on Common Problems of Genetics, Paleontology, and Systematics of the National
Research Council, set up as early as 1942. I was chairman of the committee as a whole, in absentia in most of 1942–44 while Mayr kept it
going. Dobzhansky was chairman of the section on genetics, Jepsen of
that on paleontology, and Mayr of that on systematics. In January 1947,
the committee sponsored an international conference on its subject and
in 1949 Princeton University Press published the papers presented at the
conference, edited by Jepsen, Mayr, and me. That book was another
major contribution to the development of the synthetic theory of
evolution.

Keeping on with "other things," I return to the first major research
project I took up after returning to New York in 1944. I must in fact go
back as far as 1933 to explain this. In November of that year I and my
companions were in the Chubut Valley of Patagonia looking for fossil
mammals and finding some but finding even more fossil penguins. In
fact we made the largest collection of fossil penguins that even now has
ever been made from one area and of one age. After we were back in
New York in 1934 I sorted them out and offered them to a number of
ornithologists for study. But the people who studied only recent birds
did not know how to study fossil ones and did not care to learn, and
some of the then very few people who studied fossil birds thought they
might take the penguins on, but never got around to it. So finally after
ten years during which a great many other things happened I undertook
to study them myself. The result, published in 1946, was a monograph
on our fossil penguins in particular and fossil penguins in general, and
on the origin and evolution of this peculiarly appealing group of
creatures.

That led bit by bit to something rather more than a hobby but rather
less than a whole profession. Eventually I studied and published on all
the fossil penguins known (up to the last year or two) in the world, in
museums not only in the United States but also in England, Sweden,
Argentina, South Africa, Australia, and New Zealand. In Argentina,

the Falkland Islands, Antarctica, a number of subantarctic islands, the Galápagos Islands, Africa, and New Zealand I also saw living penguins in the wild. All that activity has culminated, up to now, in a book about penguins in general and in some detail, also indulging to some extent my interest in historic voyages and in etymophily.

The next other thing to turn to involves field work and parts of my long love affair with the Southwest. Among the leading vertebrate paleontologists of the nineteenth and early twentieth centuries few collected the fossils they studied. E. D. Cope (1840-1897) was an exception. He did important exploration and collecting in the then wild West, but even he spent a small fortune hiring fossil collectors, notable among them David Baldwin, who discovered the Paleocene mammals of New Mexico. Study of those was among the things that made Cope famous. Cope's rival, O. C. Marsh (1831-1899) of Yale, led five collecting expeditions in 1870-74, the first four with students, but thereafter relied almost entirely on collectors who were paid, sometimes reluctantly. Marsh could afford that better than Cope, because he inherited a fortune from his uncle George Peabody, whose benefactions also instituted the several Peabody Museums. The one at Yale is still a major center for vertebrate paleontology. S. W. Williston (1851-1918) started as a hired collector (for Marsh) and continued collecting after he became a self-taught, but excellent, vertebrate paleontologist. (Both Brown and Granger, whose work extended well into a more recent era and whom I have mentioned before, also started as collectors, at the American Museum, and became self-taught researchers.) Scott and Osborn, the grand old men of the profession when I entered it, had collected fossils while students at Princeton but thereafter left collecting to others. Osborn did go on one later collecting expedition, to the Fayum of Egypt in 1907, but he took Granger along to do the work.

From my generation on, attitudes and procedures have changed. Practically all of the research vertebrate paleontologists have also been collectors, usually along with students or laboratory technicians or both. Indeed for most of us, decidedly including me in my halcyon period, this is the most attractive part of our profession. It takes us from offices, laboratories, and schoolrooms out into the open for hard labor, to be sure, but enjoyable labor and part of a thrilling, often rewarded hunt. It also involves travel, both domestic and foreign. I have mentioned my initiation into professional collecting in 1924, my two expeditions to Patagonia, and some jaunts in the United States. I will next tell about one of my strangest collecting trips.

Not long after I returned to the museum, a St. Louis businessman, Lee Hess, sent us some bones that had, he said, been found in the basement of a brewery in that city. I was able to identify them as being fragments of an extinct peccary, *Platygonus,* related to but quite distinct from the living peccaries, which are most characteristic of South America but range as far north as southern Arizona. (They occur here on the outskirts of Tucson and even occasionally stray into the city.) *Platygonus* is not a rare fossil. Before they became extinct more or less 10,000 years ago herds of these peccaries ranged over most of what is now the United States. The American Museum already had a complete skeleton of one. Nevertheless it was curious and worthy of some investigation that their remains should turn up in the basement of a brewery in a great city. After some further correspondence with Hess, I arranged to visit St. Louis and look into the matter further. George Whitaker, one of our technicians about whom more later, and I went there in March 1946.

It turned out that the building in question was not, or rather was no longer, a brewery and that the bones were not strictly speaking in its basement. The building had been a brewery in the nineteenth century but was now a shoe factory. While it was a brewery someone had discovered that there was a cave, an ancient channel in limestone, running under the building. It was clogged with clay and debris, but part of it adjacent to the brewery had been cleared and used as a lager where beer could be laid up at a cool and almost constant temperature. When air-conditioned storehouses made such natural lagers unnecessary, the cleared part of the cave was used for a time as a place for private theatricals and parties, then closed and virtually forgotten until Hess bought the De Menil mansion across the street from the factory. That house, long unoccupied, had belonged to Alexander De Menil, a great-great-grandson of Marie Thérèse Chouteau, the first white woman to settle in St. Louis and hailed in history as the Mother of St. Louis, although her union with Pierre Laclède, founder of St. Louis, was not sanctified by ceremony.

Hess found that the exbrewery cave probably ran under the De Menil property, had a retaining wall torn out, and discovered that indeed the cave did run on but was almost completely clogged. He set a gang to clearing it, and found the clay studded with peccary bones like raisins in a cake. Whitaker and I camped out in the grand but unfurnished and rather dilapidated mansion — George called it our puptent. With workmen provided by Hess we set up a production line, extracting the bones from the wet clay that had preserved them beautifully, taking them to our bone laundry in the kitchen of the house, scrubbing them off with stiff brushes, then treating them with preservative to prevent cracking when they dried out for the first time in some thousands of years, a delightfully simple operation in comparison with the extraction and preparation of fossils preserved in harder matrix. The whole operation and our whole stay in St. Louis were excellent fun. There were many visitors to the cave; Hess was by no means averse to publicity. A high spot occurred when someone, quite likely I, used some moderately bad language in front of a visiting clergyman, which provoked the admonition, "Profanity will get you nowhere in this cave, young man," a remark so delightfully inane that I still quote it on appropriate occasions, though rarely in caves.

Hess continued the excavation, opened an entrance to the cave on the slope below the house, put some of the bones on exhibition, and charged admission to cave and house as a tourist attraction. Years later when I visited St. Louis on a different mission I found that Hess's tourist entrance had been obliterated by a freeway and that after Hess's early death a group of civic-minded ladies had acquired the De Menil mansion, restored it, and furnished it with antiques of the period. Like Hess, they were operating it as a tourist attraction, but as a historical monument and not for personal gain. They charged a small entrance fee and also served teas on the grounds. They were most gracious to me.

In 1946 I also started a long field campaign collecting fossils in the Eocene of the San Juan Basin in northern New Mexico. In this work and thereafter until I left the American Museum, George Whitaker was usually in the field with me, and I should here briefly digress to remark on his coming to the museum. He was a young man from Valdosta, Georgia, who turned up at the museum one day and told me that he had come to work for me. I explained that the only job open was for a secretary. He said that his wife, Doris, was a secretary and would take the job. In the meantime he would work for nothing until we could pay him and thought he was worth it. Doris was indeed a capable secretary and before long we were also paying George. As I write this he is still a technician in the museum's vertebrate paleontological laboratory, although illness has curtailed his field work.

Eocene mammals had been discovered in the San Juan Basin by Cope in 1874, and, as I have already noted, David Baldwin, working for Cope, later (1880) found Paleocene mammals there. The Paleocene of that basin has become in effect a type for the Paleocene epoch and has been worked over in great detail—University of Arizona parties are spending summers there now. The Eocene had not been adequately studied and I therefore concentrated on it in 1946-54, making large collections, all, of course, with meticulous records of localities and geological levels.

We first camped at a fine spring on the western slope of San Pedro Mountain. When the young man from Valdosta first heard coyotes at night he burst out of his tent running, but he soon became as western as any native of New Mexico. The location was inconvenient and we later generally camped closer to the places where we were prospecting or quarrying. We had old exarmy jeeps, but in the rainy season, inconveniently occurring in July and August and hence in the normal field season, even jeeps could not always navigate the dirt roads and cross-country trails. The disadvantage of camping nearer our work was that the camps were usually dry, even in the rainy season, so we had to haul water, usually from a considerable distance. We tried to restrict its use for anything but drinking. I am convinced that quenching thirst is nearer to godliness than cleanliness is.

When there were no women in camp George usually did the cooking. I can't say that he has ever been completely qualified as a *cordon bleu* cook, but we survived, indeed we thrived. When his wife or mine was in camp we kindly allowed them to cook, which both we and they seemed to prefer to our doing it. One summer on the Jicarilla Apache Reservation my youngest daughter, Elizabeth, was given this privilege.

In 1950-52 we also worked on the Paleocene along the northern edge of the basin, where it extends into the Ute Indian Reservation. In 1950 and 1952 we visited my old and good friend Paul McGrew in Wyoming and then camped and collected Eocene fossils at Tabernacle Butte, in the Green River Valley just west of the Wind River Mountains. While there one summer we all came down one after the other with a mysterious ailment that we called Tabernacle Butte Misery. Anne had the worst attack and that ended the visit. As her temperature rose alarmingly I took her as fast as possible up to Jackson's Hole for treatment. A fossil locality at Tabernacle Butte is known in the profession as Misery Quarry.

Some years later we had a wonderful vacation with the McGrew's, packing with horses far up into the Wind Rivers. Come to think of it, that is the last time I was ever on a horse. My horse, Silver, did not enjoy the experience and I heard later that he was so determined not to repeat it that no one could ever ride him again.

In 1953 and 1954 George and sometimes but not all of the time I collected Eocene fossils in the Huerfano Basin, in the mountains east of the continental divide in southern Colorado. Several students worked with the parties in New Mexico and in Colorado from time to time, notably Peter Robinson, now director of the University of Colorado Museum. While visiting us in New Mexico he met the girl who became his wife, daughter of the postmaster of the town of Cuba about ten miles from our New Mexican home.

I had fallen in love with New Mexico as long ago as 1924 and when we were spending so much time there from 1945 on Anne and I began to look for a place to build a home. We had long thought the west slope of San Pedro Mountain as beautiful as any place on earth. We finally found the perfect place, about seventy acres of private land, the last up La Jara Valley, surrounded on three sides by national forest. It was wooded with grand old ponderosa pines (*pinavetes*), *piñones,* cedars (in local parlance; actually tree junipers), mountain (lance-leaved) cottonwoods, aspens, and firs. Yet from a place near the upper end one could see out right across the whole San Juan Basin and over to the Chuska Mountains on the Arizona border. From a hill just beside the place one could see north up to the snow-capped San Juans in Colorado. A visitor to the house that we built there gazed lengthily out across the Basin and finally remarked, "That would be a grand view if it was *of* anything!"

The land was owned jointly by four men, all of whom had moved away. Another man, a professor from Albuquerque, claimed that he had an option on it when he heard that we were trying to buy it, and the original deeds, which dated back to the Spanish crown, had been destroyed in a suspicious fire in the county courthouse. All these difficulties were conquered; then it took practically all one summer to get the house built, and finally in 1947 we were established there in a summer home and field base, although the actual field work was conducted from camps out in the Basin. We spent at least parts of summers there for the next twenty years and all of one winter (1951-52). Little by little, in the order named, we installed water, gas, electricity, and telephone. We had the first indoor plumbing in the valley — there was not even any in the town of Cuba when we installed ours. Gas was from a large butane tank. Electricity and telephones came to the region only toward the end of our stay there.

We got our water, clear, cool, and pure from the granite mountain, by a pipe down from the communal ditch that ran just outside our property line. When we applied for permission to pipe it in, which no one in the valley had ever thought of, the *mayordomo de la acequia* ("boss of the ditch") and some other Spanish-American neighbors came to investigate and discuss. They were adamantly opposed at first, and discussion was difficult because New Mexican Spanish is a strange dialect that I never wholly mastered. (Mexicans, whose dialect I do understand quite well, tell me that they find it hard to understand New Mexicans.) Final discussion went more or less like this:

> *I:* It is the right of every household along the ditch to take out pails of water for use in the house, is it not?
> *Mayordomo:* Yes, everyone has that right.
> *I:* I have that right too?
> *Mayordomo:* Yes, you have that right.
> *I:* Well, all I want to do is to bring in pails of water through a pipe.
> *Mayordomo:* Oh! You just want to bring in pails of water through a pipe?
> *I:* Yes.
> *Mayordomo:* Well, that's all right then.

(P.S. During the winter we spent there I had to bring in pails of water in pails.)

Here I am brought up short for a moment. In a long life there are not just hundreds or thousands but millions of episodes that seem in retrospect to be worthy of record. Choosing among them is clearly imperative but is not simple. One should not select just the formal occasions, the honorary degrees and such things. (I do not intend to dwell on any of them.) There should be a fairly good mix, with some flavor to it. How many things come to mind about the New Mexican years! There was Anne's discovery, early on, of a fine *Coryphodon* skull, lovingly prepared by the laboratory men, who sent her a photograph of it labeled Mama's First Fossil. There was the time we tied the jeep's winch cable to a stout tree to pull the car out of the mud and pulled the tree out by the roots instead. There was — but this must stop! Anne insists on just one more:

During the New Mexico years I was asked to undertake a task for the government requiring the highest security clearance. That involved FBI investigation at all my previous and present places of employment and residence, obviously including my New Mexican home. When I got back there from the field one day a local inhabitant said, "Hey, Doc. Someone who said he was from the FBI was here asking about you. But don't worry! We all told him we had never heard of you."

When we bought our Tucson house, several years before we expected to leave Cambridge, our plan was to spend winters in Tucson and summers at Los Pinavetes, as we unimaginatively called our New Mexican place. The plan did not work out. In the meantime heart problems, which we both had in different ways, became worse. We could not safely live and work at an elevation of eight thousand feet and ninety miles from suitable hospital and medical care. So in 1967 we very sadly sold Los Pinavetes. We never have and never will return. We treasure the memories of our life there, but we are happily going on in other places and with other things. It would merely be painful to go back. Some of my colleagues who used to drop in at Los Pinavetes took it as a personal affront that we were no longer there to greet them.

On to the other things. In 1947 I made my first postwar visit to Europe. The primary incentive was an invitation from the French government to take part in an international colloquium on paleontology held in Paris in March of that year. The seventeen speakers were mostly French, but Erik Stensiö was there from Stockholm, J. B. S. Haldane and D. M. S. Watson from London, Stanley Westoll from Newcastle, C. H. Waddington from Edinburgh, and I from New York. Wives were not invited, but Mrs. Haldane (Dr. Spurway) also came to Paris and participated in some of the social events. There was a distinguished group of French biologists and paleontologists, mostly from Paris but including L. Cuénot from Nancy and J. Viret from Lyon. Teilhard de Chardin, then in Paris although he soon moved to the United States, took a prominent part.

This may be a good place to remark that Father Teilhard and I were good friends for many years ending only with his death in 1955. We were indeed close friends despite the fact, well known to both of us, that we differed completely in philosophy and religion (Teilhard did not separate the two) and almost as completely on scientific matters (Teilhard never grasped the concept of natural selection and also did not distinguish science and mystical religion). Teilhard visited us at Los Pinavetes, with a companion, and we often met each other in his last years in New York. In 1971 there was a large meeting in San Francisco, arranged mainly by Joseph Alioto, then mayor of that city, for discussion of Teilhard and his work. I was the only one there who actually had known both Teilhard and his scientific work, and I pointed out that his views both on evolutionary theory and on religion were not derived from and could not follow from any of his scientific observations. The only others there who knew something about evolution, Dobzhansky and Louis Leakey, both dodged the issue—Dobzhansky by praising Teilhard for seeking a synthesis of science and religion without judging whether the synthesis was valid and Leakey by not talking about Teilhard at all. Those who knew nothing about Teilhard's scientific work generally accepted him as a great scientist.

Of course many other things were going on during this halcyon period. One was the completion and publication of the first volume of my monograph on the first three mammalian ages in South America, already mentioned in chapter 8. I should also mention that I was giving many lectures in addition to my Columbia classes. When I was in graduate school and first got up to give a paper before a scientific society I was scared half to death. I consoled myself by thinking, "This is just because I'm young and inexperienced. When I get used to this I'll be able to do it standing on my head." Alas! When I speak to an audience now I am just as shaky and frightened inwardly as when I was a graduate student, although I believe I have learned to conceal it more effectively. The details are not interesting, and I just record that I have lectured on many subjects to hundreds of audiences in most of the United States and a number of foreign countries, on lecture tours, as a visiting professor, and under many other circumstances and auspices. I wouldn't do that for money, and have not much profited by it, but I have always considered it an essential, or at least a useful, part of my profession.

I will here specify just one lecture series: the Terry Lectures, which I gave at Yale in November 1948. These lectures are then supposed to be offered to the Yale University Press for publication. I reversed the process: I wrote a book and then summarized it in the lectures. The book was *The Meaning of Evolution,* and although one or two of my books have sold more copies this one has reached the widest audience.

My account of the halcyon period will continue, but this chapter is already so long and the next thing is such a big thing that it is deferred to the next chapter.

Notes

§ It has been remarked more than once that Darwin's great work, titled in part *The Origin of Species,* was misnamed. It is about organic evolution in general or as a whole, not just speciation, which is an important part of evolution but not the only part. The founding members of the Society for the Study of Evolution were mostly systematists and geneticists used to working at the specific level, and they almost repeated Darwin's error. There is now what I consider a peculiar aberration among some taxonomists who hold that the origin of species, in the limited sense of a dichotomy yielding two species from one ancestral species, is the decisive feature to be taken into account in classification.

§ It is odd that when Osborn wrote, or had written for him, an account of the activities of his American Museum department for 1904–08, mention of the Fayum expedition was completely omitted. He did, however, write a couple of popular articles about it at the time (1907), and in his pseudoautobiography, to which I have previously referred, he credits the Fayum expedition with awakening his interest in proboscideans (elephants and their kindred). He was still working on an enormous monograph on that subject when he died.

§ The large collection of peccary bones made in St. Louis, probably from a single herd, made possible a statistical study of variation in a population, which I later published. I also published on the geology of the cave and its filling, which was more complicated than might be expected.

§ My plan for research on the Eocene fauna of New Mexico was to make a detailed revision of the whole fauna as such as soon as the collections and data sufficed. Subsequent events, culminating in my departure from the American Museum, made that impossible. Some parts of the collection have been studied but much of it has not been utilized. A successor at the museum reorganized research to cover [taxonomic] groups of fossils rather than whole faunas from particular areas and times, and those studies have not yet begun to cover all the taxa of the New Mexican Eocene fauna.

§ The account of the 1947 colloquium, published in Paris in 1950, does not list me as a "membre participant," but I was, as is evident from my remarks that appear in the publication.

References

Browning, G. O., J. L. Alioto, and S. M. Farber, eds. 1973. *Teilhard de Chardin: in quest of the perfection of man.* Rutherford, Madison, and Teaneck: Fairleigh Dickinson University Press. Pages 88–102, titled "The Divine Non Sequitur," are my contribution to the 1971 meeting in San Francisco.

Jepsen, G. L., G. G. Simpson, and E. Mayr, eds. 1949. *Genetics, paleontology, and evolution.* Princeton: Princeton University Press. This is the text of the 1947 conference at Princeton. The introduction, by Jepsen, then Sinclair Professor of Vertebrate Paleontology at Princeton University, has some details about the founding of the Society for the Study of Evolution.

Simpson, G. G. 1946. Bones in the brewery. *Natural History,* June, 1946, pp. 252–59. This is a popular account of our bone-digging under the De Menil Mansion.

————. 1946. Fossil penguins. *Bulletin of the American Museum of Natural History* 87: pp. 1–100.

————. 1949. *The meaning of evolution.* New Haven and London: Yale University Press. There are also British, Indian, and paperback editions in English, translations in ten foreign languages, and a revised edition issued in 1967.

————. 1976. *Penguins: past and present, here and there.* New Haven and London: Yale University Press.

Illustrations

I wasn't born bald but with bright red fuzz. 1902.

This is what the well-dressed young man wore shortly after we moved to Denver. Around 1904.

The whole family when I was seven. *Left to right*: Martha, Dad, Margaret, Mother, and I, still with full thatch of bright red hair.

(Left) Our Maxwell was one of the few automobiles in Denver before 1910 or so, and my father drove it far and wide. This is on the mountain road to Estes Park.

(Below) My boyhood chum, lifelong friend, and eventual brother-in-law, Bob Roe, and I early went into business together.

In the early teens, while my father was mining in Buckskin Gulch, Colorado, four of us climbed four 14,000-foot peaks in one day. Here my father, I, and Bob Roe are atop one of them. My sister Martha took the picture.

(Above) My second camp in the San Juan Basin of New Mexico, 1929. Jeff Roberson, cook and camp man at left, is responsible for the tin cans in foreground. Yes, that is a stove back of me, and an old Dodge truck back of that.

(Right) The great barranca south of Lake Colhué-Huapí in central Patagonia. We explored here in 1930 and collected four successive faunas of ancient mammals from bottom to top of the barranca.

(Above) Justino Hernández, my amiable and capable Patagonian friend and assistant, and I have found a nook out of the wind on the great barranca and we are drinking *maté*.

(Left) A still young but grizzled human and a much younger *chulengo* (baby guanaco) in central Patagonia, 1930.

(Above) Here on the 1930–31
expedition to Patagonia I am
excavating the most complete early
Eocene mammal skeleton ever found
in South America.

(Right) Back in New York in 1933 I
began the study of the fossil mammals
collected in Patagonia.

MARTY ANN

BERTHA ANN PEET   GGS DRINKING

(Above) Déjeuner sur l'herbe. On the banks of the Hudson River about 1935. *Left,* Ann (Mrs. Creighton) Peet; *middle,* Martha Simpson; *right*, Anne Roe; *background,* a paleontologist in a characteristic pose. (Photo by Creighton Peet.)

(Left) Anne and I (a bit dazed) on our wedding day, 27 May 1938.

Anne under the big tarp of our hilltop camp near San Miguel, Lara, Venezuela, in 1938. Our small sleeping tent is just visible behind the table to the left.

A meal in camp. 1938.

The gang working at the San Miguel quarry. 1938.

(Above) The Gran Sabana exploratory party at the small Indian settlement of Kamarata, southeastern Venezuela. Jimmie Angel, who discovered Angel's Falls, highest in the world, is at far left, and that is his plane behind us. His wife, Marie, is beside him. The large gentleman in the midfront was a visiting engineer, not a regular member of the party. Anne and I are to the right of the front row. The Venezuelan geologists are standing in the rear. 1939.

(Left) I am using home-made calipers on the head of a somewhat worried Kamarakoto Indian young man. 1939.

(Below) Anne and I in front of our bedroom at the Kamarata camp. That is an Explorers Club flag behind us. 1939.

(Above) The united family, about 1940. *Seated,* Joan and Elizabeth; *standing,* Helen, I, Anne, Gay (since deceased).

(Right) At war, 1944, feeling grimmer than I look. The patch is that of the Allied Forces Mediterranean Headquarters, originally under Eisenhower.

George Whitaker and I unpacking the fossil peccaries we had collected in the cave in St. Louis, 1946.

Bringing in the first important fossil of the Eocene campaign in the San Juan Basin, the skull of a primitive hoofed mammal called *Coryphodon*. Anne found it, and the laboratory men called it "Mama's first fossil." 1946.

(Right) At Oxford, England, in 1951. *Left to right*: E. B. Ford, I, Wilfrid Le Gros Clark.

(Below) Studying a mastodon jaw in Rio de Janeiro in 1954 with my coauthor of the resulting monograph, Carlos de Paula Couto.

(Below) In the pre-Andine hills outside of Mendoza, Argentina, in 1955. *Middle*, Dr. Minoprio, my collaborator there in fossil mammal collecting; *right*, Sr. Tellechea, an amateur mineralogist.

The prow of our boat, the *Ajuricaba,* on the Alto Juruá River in Brazil in 1956. The cook is preparing dinner on the forge.

Prospecting for fossils at low water along the exposed banks of the Alto Juruá River in Brazil in 1956. The boat is the *Ajuricaba,* here seen from the stern, and was our main conveyance when the water was deep enough to float it.

A gathering of American vertebrate paleontologists with the dean of British vertebrate paleontologists at Oxford, 1960. *Left to right*: Bill Clemens, Lewis Gazin, D. M. S. Watson (then professor of zoology at University College, London), Al Romer, Bryan ("Pat") Patterson, I, and Everett ("Ole") Olson.

(Right) Examining the famous Smith-Woodward autographed tablecloth kindly given to me by Lady Woodward and unwisely given by me to the American Museum. W. K. Gregory is on my right and James H. McGregor on my left; both were vertebrate paleontologists at Columbia University and the American Museum.

(Below) Our camp on the rim of Olduvai Gorge in Kenya, 1961. Philip Leakey and Griff Ewer are seated, Louis Leakey standing.

(Left) A leisure moment during a visit to the University of Washington, Seattle, in 1965. The pretty female nibbling my ear is an African bush baby, or galago.

(Below) Visiting some friends (Adélie penguins) on Torgeson Island, Antarctica. 1973.

The way we look now: Anne and I at the wedding of one of our granddaughters in 1976.

(Above) In our patio. Tucson, March 1977.

(Right) Tea at the Darwin Museum in Moscow, 14 May 1977. *Left to right:* Trofimov, Dashzeveg, Reshetov — all paleomammalogists — and I. Trofimov and Reshetov are on the staff of the Paleontological Museum of the Academy of Sciences of the U.S.S.R. Dashzeveg and Reshetov are members of the Soviet-Mongolian expedition in Mongolia.

15

Around the World in One Hundred and Thirteen Days

Ignotis errare locis, ignota videre flumina gaudebat, studio minuente laborem. ["He rejoiced to travel in unknown places, to see unknown rivers, zeal diminishing the effort."] — P. Ovidius Naso

Anne had a grandfather (I knew her family well, you recall) who would infuriate his wife by saying, "Food sure does taste good away from home." My saying is that food sure does taste good at home, and except for such special things as oysters on the half shell or boiled Maine lobster I am somewhat resistant to eating in restaurants. To that extent Anne does have a point when she says wistfully but not too seriously that I never take her out. My defense is that I have taken her, or at least that we have gone together, widely over all hemispheres and all continents of the earth and hope to continue doing so. Only once, however, have our peregrinations taken us literally around the world.

That occurred in 1951, and I think that it took just 113 days, although I may be off by one day depending on how the count is made. Crossing the International Date Line makes no difference, for the day lost there is made up an hour or so at a time before you arrive home again. The problem is whether elapsed time should count the beginning day, as the French do, making a week "eight days" and a fortnight "fifteen days," or whether to reason that a day has not elapsed until the second day, as the Americans and the English do — a difference in logic such as somehow keeps arising between speakers of English and French. I am not quite sure because I did not make a firm note of the day of departure; I think 113 days here is the French count. But enough of digressions; back to the narrative.

141

UNESCO arranged to send two Americans, two Britishers, and two continental Europeans to a celebration by the Australia and New Zealand Association for the Advancement of Science. The Americans were Dobzhansky and I; the British were C. H. Waddington and E. B. Ford; the other Europeans do not come into this account. The Americans were encouraged to go right on around the world because the air fare would be the same as for a round trip and UNESCO could thus also arrange meetings farther along the way. We could afford to pay Anne's way ourselves because *The Meaning of Evolution* had been selling well — air fares were cheaper then, but so were books and consequently author's royalties; nevertheless mine sufficed for this.

We stopped for a few days in Honolulu visiting with relatives. We had never been in Hawaii although my mother had returned there at times since her childhood there and my sisters had also spent some time there. Then on to Sydney via Canton Island, for refueling and a repair, and Fiji for another refueling. Both laps of our Pacific flight were in Boeing Strato-Clippers, which were slow, short-range, and low-flying in comparison with today's jets but incomparably more comfortable. We have never returned to Canton Island and I trust never will, but we have since spent quite a few happy days on the beautiful island of Viti Levu in Fiji. On this trip we were there only long enough for a snack in the airport (Nadi, pronounced "nahndy") in the middle of the night.

Sydney was fogged or clouded in but after we had circled for about two hours the weather began to clear. We could see a dozen race tracks here and there below us, so knew that we had well and truly reached Australia. We were soon installed in the old Australia Hotel, a place of faded but comfortable grandeur. (It no longer exists.) The afternoon was spent at the Australian Museum, a major natural history museum although not as large or as modern as several in the United States. To my surprise I was told that it was older than any still existing here. That claim was repeated to me in 1976, but in fact it is not true. The Australian Museum was founded as the Colonial Museum in 1827 and took its present name and status as a national (then still colonial) institution in 1834. In the United States the Charleston Museum used to claim that it was founded in 1773, but that claim, too, was untrue. The Academy of Natural Sciences of Philadelphia, however, despite its rather misleading name, is and has always been a natural history museum and was definitely founded in 1812. It has been in continuous although at times not flourishing existence ever since.

Our dinner that evening started with Sydney rock oysters, distinctive from ours both in flavor and in their fluted shells. Many Americans who have not been to Australia think of that fascinating continent as a vast desert or wilderness of no possible interest. We have often been asked why we have spent so much time there and we reply that we just dash over there whenever our longing for Sydney rock oysters becomes uncontrollable. I jotted down in our journal the cost of that dinner and of our room. The dinner, which included much besides the oysters and a bottle of wine, came to the equivalent of about U.S. $4 for two and the room to about $5.50 per day. In Australia that was considered very expensive. Even then it seemed extraordinarily cheap to us in comparison with New York. Things have changed both in Sydney and New York, but New York is still somewhat more expensive and very much less pleasant. Even its oysters are inferior.

Murray, professor of zoology at the University of New South Wales, remarked that "the site of Sydney is so beautiful that even the utmost efforts of its inhabitants have not sufficed to spoil it." Its setting puts it with Rio de Janeiro and Cape Town among the most beautiful ports in the world. Crossing the harbor to Taronga by boat rather than going around by the long bridge gave us an excellent view. The zoo at Taronga, on the north slope of the shore, gave us a good first look at the curious Australian fauna; in later years we were to see many of the marsupials in the wild. Like everyone else, we were especially entranced by the koalas, but surprised that these cuddly-looking creatures have fierce tempers. One that had managed to escape was caught with great difficulty. As I wrote: "They grabbed him and held him down while he struggled, scratched, tried to bite, and wailed and screamed just like a baby in a tantrum."

We were taken on an excursion into the beautiful country north of Sydney and learned, among many other things, that a female casuarina tree is a "she-oak" and a male a "bull-oak." (They do not look much more like oaks than they do like cows and bulls.) Most intriguing was an inlet where we watched an octopus changing his colors and patterns to match the background as he moved about seeking what he might devour. (They do not eat humans, but some Australian sharks do.)

From Sydney we flew to Melbourne and stayed with the Teicherts at their suburban home. Professor Teichert is a well-known invertebrate paleontologist who had migrated from Germany to Australia via Denmark. (He later moved to the United States.) A highlight of our stay was a field trip in the course of which we saw, among other things, the grooving and debris of a glacier that had moved over this now decidedly nonglacial region well over 200 million years before the last great Ice Age.

Our whole visit to Australia was elaborately and precisely scheduled, with regard for our own wishes but also for those of our hosts, and from Melbourne we were soon flown to Adelaide. There I had to give two seminars, and we met, with pleasure, all the important people in the university and scientific community. Two simultaneous dinners were given in our honor, one for the male UNESCO guests and a separate one for Anne, the only woman connected with the UNESCO delegation. Such segregation of the sexes was typical of official Australian affairs at the time; it has since been markedly relaxed, although on a later visit to Melbourne there were also segregated official dinners for Anne and me. We became rather well acquainted with Sir Douglas and Lady Mawson. He was a great Antarctic explorer whom we never saw again (he died in 1958), but some years later we did have a visit with his sprightly widow.

From Adelaide we flew to Brisbane via Sydney and spent one night there in a hotel "almost aggressively modern, air-conditioned, private bath with toilet that really flushes, electric eyes on the elevator doors — the curse somewhat taken off by the fact that one, at least, of the electric eyes does not work." The next day, or in fact the next night, we took off by taxi, having been misinformed about bus service, to Binna Burra Lodge, a resort in Lamington National Park in the McPherson Mountains, a dissected plateau near the border of New South Wales and Queensland. Here the two of us and Dobzhansky spent several delightful days. We took walks, one particularly long and varied on which Dobzhansky happily netted *Drosophila* (fruit flies, his objects of basic research) until the trail began to cross and recross a stream. For some reason Dobzhansky was neurotic about such fordings and at each, even when there was a log bridge, he stopped to take off shoes and stockings and waded across, then put them back on at the other shore, only to remove them a few minutes later. We were sore next day, but still game and happy.

"Except for the absolutely vertical cliffs (and even those have some vegetation on them) the whole region is one mass of green. A difficulty of description is that the forest is far from uniform. On our walks the composition of the forest has never been quite uniform for as much as a mile. Roughly one can speak of rain forest, open forest, and some miscellaneous types, but two parts of a continuous rain forest, for instance, may have strikingly different species of trees and make quite a different impression. Everywhere is a tremendous mixture of species — about two hundred species of forest trees are known to occur within a few miles of Binna Burra, not to mention the endless numbers of shrubs, vines, epiphytes, etc.

"Among the most peculiar rain forest trees are the strangling figs. They are planted (inadvertently) by birds on crotches or on branches of other trees. The seedling sends downward long aerial roots to the ground. Once these are established a trunk shoots up from the seedling and the aerial roots branch, increase in number and size, and grow together into a tight basketwork around the trunk of the host, which they eventually literally crush to death. Some people (including Dobzhansky, I am amused to say) view the strangling figs with fierce moral indignation — the very idea of accepting hospitality and then strangling your host. ('The tree that came to dinner'!)"

The vegetation of such a region, so unlike the stereotype of arid Australia, could fill a book and has filled many books, but must not be allowed to fill this one. So could the animal life, here mostly birds and more often heard than seen. Restraining myself, I will mention only the so-called brush turkeys, which were often seen as they came around the lodge in early morning. They do have a distant resemblance to turkeys and belong to the same order (Galliformes) but to a sharply distinct family (Megapodidae). They are best known for a very unturkeylike item of behavior: they pile up leaves, making a compost which becomes warm and saves the birds the bother of incubating eggs laid in the heap.

Thence we went back to Brisbane, boarded a Catalina flying boat, and flew out to Heron Island on the Great Barrier Reef. We were joined there by Waddington, and he, Dobzhansky, Anne, and I spent eight days (French count) there, fossicking on the reef by day and sitting in front of our cabins by night, consuming alcoholic beverages and talking. We talked of many things, from flies (*Drosophila,* of course) to the universe. All students of evolutionary theory, we were especially struck by the enormous numbers not just of individuals but of different species and higher groups (taxa in a systematic hierarchy) living together on this small island. New species usually arise when part of the population of an ancestral species becomes isolated in a geographically different area. That offered no serious interpretative problem because along the Great Barrier Reef and on in the South Pacific as a whole there are thousands of islands and shoals each of which would offer fairly effective isolation to any group of organisms that happened to reach it. The problem was the fact that in the course of time so many of the species came to be associated in the same community, or ecosystem, as the jargon has it. It is a matter of common observation in communities more fully studied that similar species live together if, and usually *only* if, each has a way of life sufficiently distinct so that they do not compete in at least one, even trivial, aspect of it. In ecological terms each has its own niche (a French word, but by English-speaking ecologists pronounced "nitch," not "neesh"). The trouble was that we could not see this differentiation for many species in the reef. Some of the different species of corals, for example, seemed to be coexisting in the same niche. We could only conclude that we did not know enough about them or that there was some hitch in the theory — also that evolutionary biology would be rather different if biologists spent more time on coral islands.

I may add, twenty-five years later, that there is now a whole mathematical theory of the control of numbers of species on islands worked out mainly by my former colleague at Harvard Ed Wilson and the late Robert Macarthur, then at Princeton. The theory works out reasonably well for some groups such as ants (Wilson's specialty) or birds (Macarthur's), but I still have doubts about the reef animals. A problem possibly even greater is posed by the large number of species of trees in a tropical rain forest, a problem that Dobzhansky also considered but did not solve.

Ever since the Great Barrier Reef was discovered by Cook, writers have been trying, with indifferent success, to describe its fascination and beauty. Of course I tried also in my journal, for the sake of our families and ourselves in later years, but of course I too failed. I will not inflict the failure on present readers. I just suggest that you go and see for yourselves.

Then back to Brisbane for the big do or bash run by ANZAAS, which took place partly at the new St. Lucia campus of the University of Queensland on the outskirts of the city and partly in various government buildings in the city. Besides the UNESCO delegates, as we were somewhat inaccurately called (our expenses were paid by UNESCO but we did not represent that organization), there was a Jubilee Delegation from the United Kingdom led by Lord Adrian. We all speechified and were speechified at and were wined and dined. I was horrified by having to reply extemporaneously to a toast to the UNESCO delegates, but apart from that and lectures that I at least had time to prepare I quite enjoyed the affair. There was one shocking incident, although it ended well. We were taken for a drive into the country, but when we stopped for a picnic lunch it was found that no one had remembered to bring a billy! (A billy is simply a tin can in which water is heated to make tea.) Although it was Sunday the proprietor of a store in a nearby village was routed out, a billy acquired from him, and we were saved from the horror of a lunch without tea.

The meeting went on for days, but on 31 May we flew by DC3 back above Binna Burra and on to Sydney where we amused ourselves and were kindly diverted by others until the evening of 2 June. Then with Dobzhansky we went on our way around the world. Qantas flew us to Cairo. (If *Qantas* was an Arabic word it could be pronounced as spelled; as it is not, everyone calls it "Quantas.") The first night we flew right across the heart of Australia to Darwin, where we were taken to an army camp for breakfast and a shower (a shower apiece, that is, and very welcome, too). Thereafter we flew only by daylight and were lodged in hotels by night, the first night in Singapore, which appalled us as a city although we enjoyed posing as international intriguers drinking gin slings at Raffles Hotel, and the second night at Karachi, which we did not see as we were lodged at the airport. For Dobzhansky and me the highlight of the trip was when we flew across the strait between Lombok and Bali in the East Indies. That strait is the site of Wallace's line, about which I will have more to say when I come to our surface visit to it years later. My journal entry there in 1951 was: "Cocktails have been produced by dear old Qantas, and we have just toasted Wallace's line with Dodick (pet name for Dobzhansky). The line did not reply to the toast." I should add, however, that the last day of the flight, which included flying right across Arabia and the Sinai Peninsula, struck me as the most spectacular of the whole journey from New York on.

At Cairo we were met by a UNESCO representative and some U.S. army medics concerned with flies and therefore interested in Dobzhansky. A bit later, at the hotel, we were joined by El Tab'i, a former student of Dobzhansky's, and a friend of his, Selim, a sociologist also educated in the United States. The name of El Tab'i has always bothered me. In classical (Koranic or *Quranic*) Arabic and all the local dialects I know of except Egyptian, one cannot say El Tab'i. It has to be Ettab'i. El Tab'i was merely confused when I hinted that he might be mispronouncing his own name. All the time we were in Cairo he and Selim were extremely helpful, beyond any possible call of duty or even of friendship. The ninth Muslim month, Ramadan, had just begun and during that month good Muslims must not eat or drink anything, not even water, between sunrise and sunset. It therefore was not merely enervating but downright torturing for the two of them to take us about in the bright sun and hot weather all day long, but they insisted on doing so. It is noteworthy, however, that we had lunch with a nominally Muslim high official who did us and himself quite well in the forbidden period, complete with wine which is forbidden at any time of day or night and in any month of the year. That was of course in his own home. No restaurants other than those in hotels for infidel tourists were open before sunset during Ramadan.

With our friends we saw the pyramids; we had Egyptian dinners (after sunset); we visited many of the mosques and the markets; we went to a village and saw a rural school as well as something of the unschooled life of the peasants (*fellahin*); we went down to the barrage of the Nile; we went to the museum and saw the treasures of Tutankhamen; we went to a night club of sorts where most of the entertainment consisted of decidedly epicene characters "dancing" by whirling in long skirts. We even did what we were supposed to be in Cairo for: we lectured (in English) at what was then Fuad I University. It was called the University of Cairo after the abolition of the monarchy. It is quite distinct from Al Azhar, the world-famous Koranic (Quranic) university, which we also visited but where we were not asked to lecture.

The sights, sounds, and smells have been described a hundred times by writers much more familiar with them, and I do not intend to vie with them. I do intend to make a few remarks about their effect on me. I had long been — and am still — greatly interested in Egypt, especially ancient Egypt. I have mentioned that I even learned to read (fairly simple) hieroglyphic texts when still in graduate school. Until it took too much time to paste it in all my books, I had a bookplate that said in Egyptian hieroglyphs "I will make thee love literature; I will make its beauty enter before thee." Of course I loved Cairo and such antiquities as I saw in and about there. Most of the Tutankhamen collection is garish and would not be valued highly if it were modern, but it is not modern. It and almost all of the exhibits in the Archaeological Museum were poorly displayed, but many truly were beautiful. What I had not expected was the impact of the relatively early mosques of the Muslim period. They are superb, in their very different way at least as impressive and beautiful as the finest medieval cathedrals in Europe — I think the mosques more so. It is a pity that the most visible of the mosques, high on a hill above the city, deserves no such superlatives. It was built by — as we say when of course we mean "for" — Muhammad Ali, not a heavyweight fighter but the founder of the dynasty still in power when I was first in Cairo. They can pull that mosque down with no complaint from me.

That reminds me that Dobzhansky, Russian Orthodox by upbringing and religious in character, was more interested in Coptic Christian churches than in mosques and spent any spare moments hunting them out in their obscurity. We went to four with him but found all of them poor things and dull despite the also honorable antiquity of a few of them.

There is something, but far less, to excite interest and praise in modern Cairo and lower Egypt. The streets of Cairo are dirty, noisy, and dangerous. As I wrote at the time, "The streets and roads are jammed with pedestrians, camels, donkeys, water buffaloes, bullock-carts and horse-carts, jeeps, Coca Cola trucks, baby carriages, bicycles and motorcycles, crawling infants, dogs, cats, and in short everything imaginable that can move or be dragged with the possible exception of reindeer sledges, and it would not really surprise me to see one of those. There seems to be a slight statistical probability that cars will pass to the right if this is convenient, but otherwise no traffic rules seem to be applied."

The disease-ridden fellahin were living in the most miserable, filthy conditions conceivable. Both in Patagonia and in Venezuela I had learned that people living in miserable conditions can nevertheless be quite happy, but I could not imagine that a fellah could be happy. Indeed an official of the Fellah Department, devoted to just this problem, agreed that the fellahin are truly miserable. There was an honest attempt being made to do something about their sanitation, health, and education, but so far with almost negligible results. I am convinced that the fellahin of 1951 B.C. were better off than those of 1951 A.D.

From Cairo Dobzhansky and we went our separate ways, we by plane to London. This was Anne's first visit to that city, and we managed to get in some sightseeing between professional affairs. The latter involved principally the British Museum (Natural History), the Zoological Society of London, and the University of London. I had old friends at all three places and managed also a brief visit with Julian Huxley, who had recently left UNESCO but was still, or even more, extremely busy. We met Lady Barlow, Darwin's granddaughter, for the first time and later visited her and her husband, Sir Alan, at their delightful country estate, Boswells. I lectured to a gratifyingly large audience at University College of the University of London and thereafter there was an official dinner given for me by the vice-chancellor.

(In British universities the vice-chancellor is the working head of a university, corresponding to the president of an American university.) I was annoyed because, exceptionally, the vice-chancellor was a woman, but she invited only males to the dinner, excluding Anne. I was less annoyed at a dinner at the zoo, which the Zoological Society operates and where it has its rooms, given by a club of members traditionally limited to men. The society as a whole is open to both sexes.

Anne was, however, included in a lunch given for us by de Beer, then director of the natural history museum, and was seated on the right of our host. Sir Gavin was his usual gracious and witty self, but he did raise Anne's hackles a bit by his low opinion of psychologists, based, as she pointed out, on the fact that British psychologists had no knowledge of people; but American psychologists did. (The British psychologists do also now, twenty-five years later.) Anne was also very much included in a wildly informal birthday party on 16 June. On one of our visits to University College I mentioned in passing that it was my birthday and Mrs. Haldane, an impromptu and hyperactive character, rushed a group of us off to a pub to celebrate.

On 18 June we went off by train to Glasgow to take part in the celebration of the 500th anniversary of the founding of the University of Glasgow. Anne wrote this part of our journal and on rereading it I am amused to see that she went into pleasant detail about our visit there but neglected to say where we were or why. She doubtless figured that we would remember well enough, and to be sure we do. Our more immediate hosts, the departments of zoology and of geology in the university, had been misinformed about our arrival and when we arrived in the evening, tired and hungry, we found that a reception for us had been under way for some time and they would be happy if we would attend.

The several days of the celebration were full of ceremony and pageantry. The delegates from universities and learned societies and other invited guests wore robes for the ceremonies and processions and made a noble show. Costumes were of every imaginable color, and some in combinations we had not imagined. Most impressive, perhaps, were the delegates from some continental university (I don't remember which) in a black and red robe with a medieval ruff and the delegate from the French Academy in black tail suit embroidered in green and gold, wearing also a cocked hat and a sword. At the two evening receptions the men wore formal dress suits and robes over them. The women, on the whole somewhat dowdy in their informal wear (Anne tells me), blossomed into true splendor for the formal evenings. At one of the receptions, the one given by the lord provost and magistrates of the city, part of the entertainment was singing by a girls' teen-age choir. They sang very well, and while doing so all took on the same expression: mouths wide open, eyes slightly popped, staring fixedly at nothing. Ever since then when someone looks similarly inane we call it "the Glasgow look."

The last day, 21 June, was to be the climax because the king and queen were to be there. But, the king was ill and the queen did not want to leave him, so they sent Attlee, then prime minister, in their stead. (Some months later King George died of that illness.) Attlee gave a speech that was not informative but was so elegantly phrased and so gracefully delivered that even hardened Tories, angered by the nonappearance of royalty, applauded it loudly. Later, precisely on schedule at 10:40 P.M., Glasgow University began its second 500 years.

Then by train on to Edinburgh, where we spent a couple of days with Wad — C. H. Waddington — and his wife Justin in an eighteenth-century mansion in the country divided into flats where Wad's staff lived. We saw something of Edinburgh, a charming city in some ways but not near the top of our list of such, and also of the surrounding region, also charming. Back to London to catch up on mail and for visits, notably with D. M. S. Watson and Lady Woodward. Watson, an eminent vertebrate paleontologist, had just recovered from a serious illness and was about to retire from University College. His successor as professor of zoology, Medawar, is not a paleontologist. Lady Woodward was the widow of Arthur Smith Woodward, another eminent vertebrate paleontologist of an even earlier generation (1864-1944). Sir Arthur had already retired when I first went to England, but he was hospitable to a youthful acolyte. On this later visit Lady Woodward gave me a tablecloth on which she had embroidered the signatures of the many late nineteenth- and early twentieth-century scientists who had visited the Woodwards. Feeling that this was a sort of relic too precious to be in personal hands I later gave it, quite unwisely, to the American Museum.

Thence northward again by train to Newcastle, where we were met by an old friend, Stanley Westoll, a vertebrate paleontologist, and stayed with Lord Eustace Percy, younger brother of the previous and uncle of the then present Duke of Northumberland. Lord Eustace, after a varied career, was then rector of King's College and pro-vice-chancellor of the University of Durham. King's College was not a college in either the British or the American sense but was a branch in Newcastle of the University of Durham, in nearby Durham. (King's College later became the separate University of Newcastle.) We spent some time in Stanley's department and in a local natural history museum, but had come for ceremonies at Durham, which wound up in the cathedral there, the greatest extant masterpiece of Norman architecture, late eleventh and early twelfth century. That style and its French equivalent (Romanesque) have fascinated me ever since I first saw an example of it, and I thought Durham Cathedral grander than Vézelay, the Romanesque masterpiece. Parts of the more modest college buildings are older and also fine.

Then back to London by train and to Paris by plane — if you think we were moving about a great deal, you are quite right. We had a lovely time in Paris, at the natural history museum, visiting vertebrate paleontologists and anthropologists, sightseeing, eating at good restaurants, and visiting the Belpaumes, with whom I had been billeted in Algiers for a time during the war. A typical event was what Mme. Arambourg, wife of Camille Arambourg, then head of paleontology at the national museum, called "an insignificant family lunch," which consisted of seven courses with four wines followed by coffee and liqueurs and went on for more than three hours. And an incident of which Anne still reminds me was buying her a fancy blouse at a chic shop, where a beautiful oriental sales lady enticed me into the trying-on quarters by saying, "Come wiz me!" — and I went.

I had been ill off and on since leaving the United States and I became rather badly so in Paris, as a consequence of which we stayed there longer than intended. I was attended by Bourlière, a well-known mammalogist who is also a doctor of medicine, and by 17 July we were able to get on to Zurich. We stayed there with Frau Peyer, wife of a vertebrate paleontologist who happened then to be away in the United States. We made an excursion into the Alps and also one to Basel to see Drs. Hürzeler and Schaub, two more vertebrate paleontologists.

From there we visited Belgium, new to us both: the museum in Brussels, the zoo in Antwerp, and the sights in Bruges. To London, and up to Oxford where we stayed with Wilfrid Le Gros Clark, an old friend then professor of anatomy there, and his first wife. (She predeceased him and he then married an old family friend.) The vice-chancellor gave us a lunch there and as at the director's lunch in London Anne sat on the host's right and again Anne, who is always polite (well, as nearly always as is humanly possible) and always (yes, always) frank had occasion to set him right. This time it was his sociological views: aristocratic and antifeminist.

Our tour of the world was beginning to seem endless to us, and perhaps is beginning to seem so to the reader, but it had one more section and that one of the best: our first visits to Norway and Sweden. In Oslo we visited with Anatol Heintz, another — you guessed it! — vertebrate paleontologist, and did the usual tourist things. By train and bus we went to the Norwegian Academy of Science's farm in the central Norwegian mountains, with some difficulty because we spoke no Norwegian and, this being off the tourist track, no one along the way spoke anything else. The academicians visiting the farm did speak English, French, or both, and we had a splendid time there with them and in the alpine scenery. Although the altitude was not great we were far enough north to be only an easy climb below timberline, which made us former Coloradoans feel quite at home. We also learned to enjoy gjetost, which we had scorned before as poor cheese but liked when we learned that it is anticheese, made from whey and not curds. (We still love good cheese made from curds.)

In Sweden we did not find the ambience and some people as simpáticos as those of Norway, but here, too, were friendly and thoroughly simpáticos colleagues, notably Stensiö, now the grand old man of vertebrate paleontology — I am happy to say that at this writing he is still alive and active at the age of eighty-five. It was crayfish season, which is a big time in Stockholm, and in addition to the usual museum visits and some sightseeing we had a remarkable crayfish-plus dinner with Stensiö. We also went to Uppsala for a day and visited the university and the house and grave of Linnaeus.

And finally back home. Anne made the last entry in our journal. I quote:

> Conversation in waiting room at Stockholm airport — all in loud British voices — between Sir Archibald and Lady.

> *Lady* (heading for toilet): I think I'll just go here.
> *Sir A.:* Ah, be as quick as you can, won't you?
> *Lady:* Oh, well, I won't go if you'd rather —
> [Pause.]
> *Sir A.:* Don't you want to go?
> *Lady:* All right. Call me, won't you?

Notes

§ I still regret the comfort of some of the old prop planes, which in this respect were mostly superior to first class on a 747. The army C47 was almost unbelievably uncomfortable and was only slightly less so in its civilian incarnation as the DC3, but the latter had the fully compensating virtue that you could see the scenery. So you could to less but still considerable extent in the Strato-Clippers and the Constellations. Now miles higher it is rarely worthwhile to look out, and few passengers ever do.

§ Although Australians of professional stamp often speak British English, or try to, the real Australian speech is different from either British or American English and often hard for an outsider to grasp. It is not simply cockney, as some travelers have insisted. *Fossicking on the reef,* as we did on Heron Island, is good Australian. To fossick means to rummage about or go prospecting.

§ The archetypal British Association for the Advancement of Science, among many other things famed for the debate on evolution between T. H. Huxley and Bishop Wilberforce, is widely known as the British Ass. The name of the corresponding Australian and New Zealand association is generally pronounced "Anz Ass," perhaps without any ulterior meaning.

§ Arabic abounds in sounds that do not occur in European languages and there is no satisfactory or simple way to write Arabic words with our alphabet. For example, the sound we write as *d* in *Ramadan* is different from that of the Arabic letter really pronounced *d* and is not pronounced as in English. The frequent English spelling *Moslem,* however, merely represents an unnecessary mispronunciation of the real Arabic word, fairly well represented by *Muslim.*

§ Perhaps I should add that on this trip I received my first non-American honorary degrees, Sc.D. from both Durham and Oxford and LL.D. from Glasgow. It is odd for a scientist to be a (nonpracticing) Doctor of Laws, but for some reason the honorary degrees at Glasgow were only that and Doctor of Divinity, and to some people the latter would seem still less appropriate in my case or even blasphemous.

References

Macarthur, R. H., and E. O. Wilson. 1967. *The theory of island biogeography.* Princeton: Princeton University Press. Tough going, but having mentioned it in the text I thought I should cite it.

Simpson, G. G. 1942. The first natural history museum in America. *Science* 96: pp. 261–63. I intended this short article to settle the matter, and apparently it did as no one has contradicted it; this is the answer to the claim of the Australian Museum.

Waddington, C.H. 1975. *The evolution of an evolutionist.* Ithaca: Cornell University Press. The title suggests an autobiography, but the book consists almost entirely of reprints of articles previously published elsewhere. Little of it is directly autobiographical, but it does illustrate Waddington's mental development and his rather insistent claim to fame. My name does not appear in the index, but it does twice in the text (pages 89 and 280); in both places Waddington commenting on a short paper by me that he does not seem to understand.

16

A Drop of Nature

"There's nothink so loverly as a loverly drop o' Nachure." — Margery Allingham, in the *persona* of Mr Lugg

Mr Lugg, a sophisticated character despite his somewhat informal speech and manners, well expressed an opinion I share. In the several years following our trip around the world I reveled in the loverly aspects of Nachure, and met with pain but no regrets an exemplification of the fact that she also has a grim, even deadly aspect.

After returning from the trip related in the last chapter and catching up a bit in New York, we moved to our New Mexican home, Los Pinavetes, with a quantity of publications and notes. There we spent most of the following year. The winter is severe at that elevation, but we were snowed in for only about two weeks and the rest of the time could usually get down by jeep the five miles of dirt (or mud, or ice) road to post office and store. Even when snowed in we lived well because, as Anne likes to tell, we had sourdough working on a mantelpiece, a case of gin in a closet, and half a hog hanging in a storeroom. Much of the time we slept in the living room, which we did not attempt to heat, and worked in the bedroom, smaller and easier to heat.

Our work, for which the calm and isolation were ideal, was mainly writing. Anne wrote *The Making of a Scientist,* a popular book summarizing the results of long research on the personalities and life histories of eminent scientists. My first task was to complete writing *The Major Features of Evolution,* which had started out as a revision of *Tempo and Mode in Evolution* but had turned into an essentially new work. That was out of the way by the end of the year, and I then turned to writing a fairly short and not unduly technical book on general paleontology.

This had been commissioned by a British publisher, but when I sent that firm the manuscript they returned it with a curt statement that it was not exactly the sort of thing they had in mind. I was daunted, but I tried it on Yale University Press, where they decided that it was exactly the sort of thing they wanted.

That was *Life of the Past,* and I hope the British firm that refused it knows that besides hardback it went into two paperback editions, one put out by Yale and one by Bantam Books, and was also published in Spanish and in German. It had some good reviews, and one bad one by a British zoologist who objected violently to the illustrations, which I had drawn myself during that winter at Los Pinavetes. I admit that my drawings are crude and inartistic, but they have a certain amateur freedom that some people find attractive or at least amusing. What did annoy me a bit was that my critic had also illustrated some of his publications and that his drawings were just as crude and inartistic as mine, and moreover that he had the poor taste to die before I could point that out to him.

The potential importance of that book was that up to that time there had been no book in English that was truly on general paleontology, that is, the background, the principles, the aims, and the significance of paleontology as distinct from mere systematic description of the various groups of fossils. There had been one such work in German, published by Walther in 1927, but I was not aware of it and it had not been translated. Years after my book appeared, a general paleontology (under that little) was written by A. Brouwer, published first in Dutch in 1959, and in English translation in 1967. Although Brouwer cited several of my other books, he did not mention the one that antedated his in the same field. Finally in 1971 appeared a fine textbook by Raup and Stanley, written in English at length and at a technical level, that surpasses previous efforts, including my shorter, less technical, and now obsolete work. They, too, cited other books by me but not this one. Thus in spite of its rather wide circulation, I cannot maintain that *Life of the Past* started the trend to consider paleontology as a general subject and not only as a complex of taxonomic specialties. Incidentally, Raup and Stanley also did not cite Brouwer's previous book on their subject or the historically important work by Walther.

We worked steadily, but wintering at Los Pinavetes was not all work. The evergreen forest was beautiful in the snow. Also while snow was on the ground (most of the time) each morning we were entertained by the trails that appeared in it. Few animals really hibernate all winter, and tracks of many species were made overnight. Perhaps the most interesting was a double bear track, one set of prints made by a large animal, the other set by a relatively tiny cub. A large flock of wild turkeys moved down from the mountain and roosted in the ponderosa pines at the top of our property. The Forest Service gave us corn to feed them, but at first they ignored it, apparently not recognizing that this strange stuff was food. Mobs of Steller's jays moved in to clean it up — they recognize anything as food. Finally one day an adventurous turkey tried a grain or two, the others saw what she was up to, and all came running back as fast as they could to get in on the free supper. After that they fed regularly and we moved the food progressively closer to the house until we had them eating just outside our window at the cocktail hour. Come spring we had a startling sight: one day here came a big tom strutting in full display, leading a parade of his harem.

There were also some social events. When the road was open we occasionally had house guests, and there were sometimes parties in La Jara (our post office, a store, a bar, and a few houses) or Cuba (five miles farther away, another post office, more stores, more bars, some more houses). We gave a dance with an old-time fiddler, babies stashed in the bedroom (as in the famous episode in *The Virginian*), and a new bottle (not for babies) appearing mysteriously in the dining room whenever one was empty.

I voted that November at the La Jara schoolhouse, laboriously built by the valley people themselves. School there was illegally taught partly in the local kind of Spanish. When I went to vote one poll watcher asked another in Spanish, which he did not think I understood, "Hey! Can this guy vote?" Reply: "Yes. It's all right. He's a Democrat." After I had marked my ballot, it was carefully examined to make sure that I had indeed voted a straight Democratic ticket before it was put in the box. The only Republican in the valley was the Anglo storekeeper. Somehow he got elected as a state senator, certainly not by the La Jara Valley vote. My vote did us some good because the Democratic road commissioner, in private life a barkeeper in Cuba, had the maintainer plow out our private road from time to time, especially when we were giving a party to which he was invited.

In summers I continued to collect Eocene fossils. The two winters after that one were spent in New York, living in hotels because the girls were away at school and we had not kept an apartment when we went on our long jaunt around the world. Besides museum routine and teaching for Columbia I did some office research, although none of any great importance in this period. I also did some lecturing at other universities, notably at the University of Oregon and Oregon State, where I gave the Condon Lectures in January 1953, on the subject of historical biogeography.

In 1954 the Brazilian National Research Council invited us to visit that country to lecture, do research, and consult with Brazilian scientists. By that year it was possible to fly down to Rio, but we preferred to go by ship, which we did with stops at Barbados and at Baía (modernized spelling of Bahia) in Brazil. On Barbados we learned that flying fish are delicious, although it does seem a shame to eat those delightful creatures. At Baía we went to a maternity hospital, not for the usual reason but because Anne had broken one of the tiny bones (a sesamoid) in a toe and the maternity hospital had the only x-ray apparatus. We also were treated to lunch by friendly Brazilian fellow passengers at a posh hotel where the dish of the day was truly exotic, for Baía that is: Irish stew. (During a later trip to Brazil, at a restaurant in Manaos, which is situated on a river teeming with excellent fish, the dish of the day was salt cod imported from the U.S.A.) We had hoped for something a bit more Brazilian, and our host created a furore by insisting that we be served what the cooks and waiters were themselves eating. That turned out to be vatapá, a delicious dish that we never again encountered. The cuisine of Baía is different from that of any other part of Brazil; among other things it is more spicy.

In Rio we were met by two Brazilian vertebrate paleontologists whom we already knew and with whom we would principally be working: Carlos de Paula Couto and Llewellyn Price. Llew, in spite of the name bestowed on him by American parents, is a native Brazilian. Both he and Carlos were born in the town of Santa Maria, which happens to be a famous fossil locality, in the state of Rio Grande do Sul. At that time and for long after Carlos was on the staff of the Museu Nacional, in the former palace of Dom Pedro Segundo, once emperor of Brazil, and Llew on that of the federal Divisão Nacional de Produção Mineral (DNPM), equivalent to a geological survey, almost at the opposite end of the city. Both institutions have noteworthy fossil collections, and Carlos generally studied the fossil mammals and Llew the fossil reptiles in both. When I arrived the most interesting fossil mammals were a large collection of mastodonts that had been made by Llew, and it was agreed that Carlos and I would write a joint monograph on them. I was given a large office in the DNPM and Carlos and I were soon at work there. Before long Anne often worked there too, working over a draft of her book *The Psychology of Occupations*.

Carlos had found lodging for us in a *pensão* (a somewhat flossy term for a boarding house) in Copacabana, which is not only a splendid beach and waterfront but also a whole borough (*bairro*) of Rio. We were quite comfortable there, although like all *Cariocas* (inhabitants of Rio) we were often reminded of the popular jingle:

> *Rio de Janeiro*
> *A cidade que me seduz,*
> *De dia não tem água,*
> *De noite não tem luz.*

> [Rio de Janeiro,
> the city that enchants me,
> by day it has no water,
> at night it has no light.]

— And that reminds me of our struggle with the Portuguese language. On the ship we had spent several hours a day studying Portuguese, and we continued to study it in Rio. One reason for living in a *pensão* was that it forced us to speak Portuguese. (Other reasons were that we wanted to get acquainted with ordinary life in Brazil and that we had a sufficient but fixed stipend for expenses that would not cover long stays in tourist hotels.) Some people who know only one of the languages think that Portuguese and Spanish are so similar as to be almost interchangeable. That is not true. Indeed we found our previous knowledge of Spanish to be something of a stumbling block in learning Portuguese. Even similar words in the two languages are often pronounced and used differently. For example, both have related words frequently translatable as "to have," *tener* and *haber* in Spanish, *ter* and *haver* in Portuguese, but they are conjugated and used quite differently in the two. It is really easier when synonyms in the two are of entirely different etymology, for example *sombrero,* Spanish, *chapeu,* Portuguese, *hat,* English. Portuguese is far more difficult than Spanish, more complex both in grammar and in phonetics. We did not exactly triumph over those difficulties, but we did cope with them after a fashion. Before too long both Anne and I were giving lectures in Portuguese of a sort, not the most impeccable sort, to be sure, but still comprehensible to our Brazilian audiences.

We enjoyed our fellow boarders and soon became good friends with them. Among them was a very likable but sometimes confused old lady. One evening at dinner we were all talking about UFO's, of which there was a plethora reported in the Brazilian newspapers. The others all agreed that the flying saucers are indeed visitors from outer space, but Anne and I expressed doubts. The old lady put an end to that by saying that she had often seen them and had spoken to the occupants. They said they came from Mars.

That nice old lady had a rather dark complexion — "café com leite," in Carioca terms. Her children were somewhat lighter. There is no firm racial line in Brazil and white grades imperceptibly into black without division or stigma, although it is true that the richer Brazilians are, on an average, whiter than the poorer ones. There is a Carioca saying that "Prêto rico é branco" "A rich Black is white."

In late November Carlos and we, in furtherance of the study of mastodonts, flew to Belo Horizonte, capital of the state of Minas Geraes. Near there are some caves, *lapas,* from which many Pleistocene mammals had been collected, first by a Dane named Lund who went there in 1834 and spent the rest of his life there. We visited the caves, but before that, upon our arrival, there was some confusion at the university when Carlos announced that I had come to give a lecture, which was complete news to me and to the university officials, who had never heard of me. That was tactfully straightened out (no lecture) and we did examine some fossils at the university and some others, more connected with my current research, in the (illegal) private collection of the British vice-consul, an amateur paleontologist who had been grubbing in the lapas. Carlos tried unsuccessfully to get him to donate the specimens to the Museu Nacional. In the process he used a charming Brazilian proverb: "Quem não chora não mama" — "Who doesn't cry doesn't suckle" — that is, if you don't make a fuss you don't get what you want. (I just cannot get entirely off the subject of language: there are many orchards around Belo Horizonte and I learned that the unit of fruit trees is *pé,* "foot" — "ten feet of oranges" is not a string of oranges laid out to that length but ten orange trees.)

Anne became quite ill in Belo Horizonte, but she was helped on her way to a quick recovery when the faculty of the university sent her an orchid corsage and I bought her some very good, amazingly inexpensive gems: a large aquamarine, a large amethyst, and a "play" of citrine or quartz topaz, two stones for earrings and one larger for a ring. (I later had them beautifully mounted in Buenos Aires.) Minas Geraes is the source of an extraordinary number of gems, ranging all the way from emeralds to agates.

Thence off by car and hedge-hopper small plane to Araxá, passing on the way a modernistic church that had never been consecrated because the relevant bishop objected to a bas relief of a very explicit, very seductive, entirely nude Eve.

Araxá, pronounced quite approximately ah-ruh-*shah,* was a main objective because the large collection of mastodonts made by Llew Price came from there and some specimens were still there. There are also springs and the government had built a huge — under the circumstances I should say "mastodonic" — establishment: a luxury hotel to accommodate some seventeen hundred guests, a hydrotherapeutic establishment, and a casino, closed because gambling had recently been forbidden by a law passed by the same government that had built the casino. We occupied two large suites, the most luxurious field quarters that I or, I feel sure, any other vertebrate paleontologist ever had. Just eighty-two other guests rattled around in this amazing establishment.

In front of the hotel was a free-form structure with a bath house, two ramps down to spouts of fetid warm spring water, and a pit containing a mass of mastodont remains *in situ.* Inveterate builders, the Brazilians had gone so far and no farther. There was not even a sign saying "mastodont bones." A guest at the hotel said he had been trying for days to find out why an enclosure had been put around that mess of rubble and no one could tell him. While Carlos and I were studying the specimens Carlos explained to onlookers that mastodonts originated in Africa and walked thence all the way across Asia and North America down here to Araxá. A young lady exclaimed. "No wonder they died when they got here after that long walk!"

We hedge-hopped from Araxá to São Paulo on the aerial equivalent of a milk train and were met at the São Paulo airport by geneticists Pavan and Brito da Cunha and also to our great surprise by our Egyptian friend El Tab'i. His mission here was also somewhat peculiar for a Muslim, supposedly teetotal. Studying *Drosophila,* he had become an expert on yeasts, food for *Drosophila.* The national drink of Brazil, distilled from yeast-fermented cane juice and variously called *cachaça, caninha,* or *pinga,* generally has an unpleasant odor (unpleasant taste, too, according to Anne). El Tab'i had been employed to try to breed a yeast that would not produce such an odor. Later in his laboratory we tried samples of the cachaça so produced, and even Anne found it quite palatable along about the tenth or eleventh sample.

There was no really interesting paleontology for us there, but some interesting genetics, geology, and sightseeing. The geology nearby in this state is especially interesting for its comparison with that of South Africa and its bearing on continental drift in premammalian or only earliest mammalian times. On that field excursion we were also interested in the town of Americana, founded by diehard Confederate soldiers at the end of the Civil War. Their descendants now are completely Brazilian and have a mixture of Negro blood.

In São Paulo I was expected to lecture and did so to an overflow but I fear not overenthusiastic audience. We also had a visit with Vanzolini, a herpetologist who had studied at Harvard and was also a TV performer and popular music composer. I noted that the only words I remembered from one of his songs were:

> *Bebendo com outras mulheres . . .*
> *. . . Jogando bilhares . . .*
>
> [Drinking with other women . . .
> playing billiards . . .]

— which is a false rhyme, and a somewhat gooey bit about a *moça* ("maid") objecting to the behavior of a *rapaz* (a flip male teenager). He is a good herpetologist.

Among other amusing events we went to a birthday party, where I acquired one of my extremely few party tricks, singing "Happy Birthday" in Portuguese:

> *Parabéns a você*
> *Nesta data querida.*
> *Muitas felicidades,*
> *Muitos anos de vida.*

I also heard one of the very few Brazilian jokes that I think is really funny, and I cannot withhold it although it has to be in Portuguese:

> Pai, numa loja fonográfica: *O menino quer um disco.*
> Empregado: *Um disco infantil?*
> Pai: *Não é. Um disco voador.*

(See following note for translation, but the point is lost in English.)

For some reason that reminds me of one of the most backhanded compliments I ever received: "You speak Portuguese so well that I thought you were German!"

Back in Rio, by Christmas I had about finished work on the Brazilian mastodonts and Anne had almost finished the final typing of her book, so we took the holiday off and went up to Petrópolis for a few days. The weather was hot — Rio is in the tropics, although only barely so, and Christmas comes at the hottest season there. Petrópolis was established as a summer resort by Emperor Pedro II, for whom it is named. It is well up beyond the great scarp of the Serra do Mar and is relatively cool. We found it relaxing, and it was also a treat to have a private bath and to be in a small, well-run hotel where there was *água* and *luz* both *de dia* and *de noite*.

We had a couple more hectic days in Rio but then started traveling again. Simple itinerary: Curitiba, capital of the state of Paraná; more lecturing and a geological field trip. Foz do Iguaçu baggage did not arrive and accommodations were difficult, but the Iguaçu Falls, which are just there near the junction of the Iguaçu and Paraná Rivers and of Paraguay, Brazil, and Argentina, are one of the grandest sights on earth. Back to the Grande Hotel Moderno in Curitiba, our favorite Brazilian hotel, largely because it was not in fact *moderno*. Pôrto Alegre, capital of Rio Grande do Sul, southernmost state of Brazil; more lecturing (always in Portuguese now and a terrible strain, even worse than in English). A difficult but worthwhile excursion to Santa Maria. Back to Pôrto Alegre. Buenos Aires.

We arrived in Buenos Aires on 15 January and were greeted there by Bryan Patterson, an old friend who had been studying Argentine fossil mammals but was soon to return to the United States, and Noemí Cattoi, then in charge of vertebrate fossils at the museum in Buenos Aires. For ten days Anne became acquainted with Buenos Aires, which she had not previously visited, and I worked as hard and fast as I could in the great museums in Buenos Aires and La Plata (then temporarily called "Eva Perón"), catching up on some needed additional observations for my big, half-published work on the early Cenozoic faunas of Patagonia and for the monograph on mastodons. Then on the twenty-sixth we flew on the internal Argentine air line (very uncomfortable and badly managed) to Mendoza, a city in western Argentina almost at the foot of the highest part of the Andes. That is the metropolis of the main wine-growing area of Argentina, or of all South America, and very good some of its vintages are, even though we somewhat prefer a couple from Chile and, oddly enough, one from northern Patagonia. That reminds me of a country song some of my Patagonia pals used to sing in what they claimed was a Mendocino accent, going (in small part):

> *E' que así semo'*
> *Lo' Mendocino'*
> *Como uva que dan la' parra'*
> *Cuando más se las estruja,*
> *Más jugo largá.*

[That's the way we are,
the men of Mendoza,
like grapes from the vine,
the more you press them,
the more juice they give.]

(The idea is that the tough guys give as well as they get.)

Although it is a fairly large city, Mendoza retains a somewhat rural feeling with its low profile and tree-shaded streets with open ditches on the sides. During a couple of visits it became decidedly our favorite Argentine city.

Here I had a pen pal, J. Luis Minoprio, a medical doctor and an amateur paleontologist. We had a double mission in visiting him, in addition to the pleasure of meeting him and his family. Anne had been suffering for years and I more briefly from chronic brucellosis (undulant fever), and our New York doctor (the same one who played in the Mandolin and Martini Club) had been independently in touch with Dr. Minoprio, who had developed a special diagnostic technique for that affliction. Minoprio tested us, and to our dismay the result was strongly positive. (We did later recover under new treatment.) Minoprio had also been sending me some fossils from a quite peculiar mammalian fauna found just on the outskirts of Mendoza. While there I saw some additional specimens and also went over the exposures from which they came.

From Mendoza we also took a glorious excursion into the high Andes, up to Uspallata Pass, at the border with Chile, about 12,500 feet in elevation and with a fine view of Aconcagua, at 22,835 feet the highest mountain in the Western Hemisphere.

On 2 February we returned to Buenos Aires, and although we were both extremely fatigued, we flew the next day to Mar del Plata, a fast-growing summer resort town on the Atlantic shore in southern Buenos Aires Province. (The city of Buenos Aires is not in that province; it is a separate Federal District.) We went there, most inconveniently when at the height of summer the city was crammed by some 2 million visitors, in order to visit the local museum that I have mentioned before. We were greeted by its director, Galileo Scaglia, and also by Jorge Lucas Kraglievich, son of the Lucas Kraglievich mentioned in connection with my stay in Buenos Aires in 1931. In 1955 the museum there was still jammed into one room in the city hall or municipal building, but when we again visited it some years later it had its own attractive building in a park on the waterfront. The purpose of my visit in 1955, soon accomplished with the always cordial assistance of Scaglia, was to pick out specimens useful for my then current studies of South American fossils and arrange to have them sent on loan to me in New York.

We also took the opportunity to visit the famous fossil localities in the vicinity, and that made the occasion for a fairly typical picnic lunch for visiting colleagues. As I haven't before, I'll give the menu as an example:

Apéritif: vermouth
Cold boiled shrimps
Clams grilled in the shells   } White wine
Beefsteak
Matambre (another form of beef) } Red wine
Miscellaneous sweets

(With good bread but no salad or vegetable — my kind of meal.)

We also took a five-day excursion with Scaglia, Kraglievich, and Reig (another Argentine vertebrate paleontologist), visiting other fossil localities along the southern shore of Buenos Aires Province. After brief stays in Buenos Aires and Rio to wind up things and see to shipping baggage, we flew back to New York, where we arrived on 20 February.

Now I am going to defer discussion of some intervening developments in the United States and skip to my next visit to Brazil, in some respects among the best of my many travels, in one respect much the worst. In the 1950s the Cenozoic mammalian sequence was reasonably well known for Argentina in the south and the latter part of it for Colombia in the north, but there was very little knowledge of the vast area in between. A few scraps had come from the Purús and the Juruá, large rivers tributary to the Amazon from the southwest, and it appeared that toward the headwaters of one of those rivers would be an appropriate place to look for fossils. An arrangement was made for a joint Brazilian-North American expedition, and we decided on the Alto ("Upper") Juruá starting from Cruzeiro do Sul ("Southern Cross"), effectively the last outpost of civilization in that direction. The paleontological-geological-mapping party would consist of Llew Price, a North American graduate student, George Whitaker as my field assistant, and me. The Brazilians also sent a party to act somewhat independently and to collect recent birds and mammals.

We had to aim for the so-called dry season, when it rains only half the time instead of all the time and the river is low. The only visible geology and possible fossil-collecting places are on shoals and river banks, and in the wet season the river rises some forty feet, overflows its banks, and makes everything but the dense rain forest invisible. The North American contingent took off by plane on 27 May 1965 and landed at Belem early the next morning. By hard stages we eventually got to Cruzeiro do Sul on 3 June, the last stage in a beaten-up Catalina, an amphibious plane that could use the river as a landing strip, although near Cruzeiro there was a dangerously short land strip cleared in the forest. The trip to Cruzeiro included a delay along the way in Manaus, a rubber-boom city that had gone bust after the boom. It was called by one of our Brazilian companions *"a terra de teve,"* which cannot be so neatly expressed in English but is roughly "the land of used-to-be." I found it "a town like none other on earth, and well worth the price of admission, whatever that turns out to be." It turned out to be very high indeed.

Eventually we reached Cruzeiro do Sul, which was (doubtless still is) a considerable town for so remote a spot. Widely scattered houses. No streets but trails all over, some in the center of town paved. A town hall (*prefeitura*), post office, bank, a square of shops, a market. A few horses but most loads carried by humans or by small, wiry bulls. No plumbing but potable well water. Electricity from 6 to 9 P.M. when the motor worked. We were there a long, weary time because the boat originally hired was stuck upriver because of low water, another engaged in town needed repair, and our outboard motors were long delayed. We were kindly taken in by Dona Raimunda Ruela, who had a large but already somewhat crowded house. It was beneath her dignity to rent us rooms, but she decided she could accept us as house guests if we arranged to eat elsewhere except for morning coffee and snacks with which she plied us all day. A nice woman, talkative and a booster for her home town: "Those people down in Manaus think we are savages up here, nothing but forest and jaguars. Why! Jaguars rarely come into town. This is the healthiest place in Brazil. Almost no tuberculosis and only a few dozen lepers. The malaria is not bad this year. This is real white man's country. It takes a little planning to get food, is all."

Cruzeiro is the head of navigation for the stern-paddle steamers that then plied the Amazon and its tributaries. One got that far while we were there even with the river low, but it hurriedly unloaded and got out lest it be caught there idle until next high water. We passed some time making short excursions in small boats from Cruzeiro up the Juruá and a relatively small tributary, the Rio Moa. Even on these first tentative contacts the greatest of the earth's forests had an impact that was only to deepen with the broader acquaintance of following weeks.

"A feeling of utter astonishment at the prodigality of a great tropical forest — prodigality of everything, species, individuals, a piling up of life in incredible depth and complexity. One or two hundred species of tall trees form a framework, which is then wound about and almost shrouded in vines, and then every crotch and branch where a little sun penetrates beneath the canopy becomes crowded with epiphytes, here mostly the varied bromeliads. As always in such a forest, animal life is remarkably scanty or, rather, remarkably invisible. There are certainly many creatures everywhere, but you just don't see most of them. The water, muddy even at this low stage, conceals its millions of fishes, dreaded stingrays among them, turtles, alligatorlike jacarés, and water snakes — only the dolphins do come up regularly, blow very quickly, and roll back under again. Butterflies are much in evidence at the edge of the forest and on damp spots along the bank, notably the wonderful big iridescent blue Morphos so associated with thoughts of the Amazon Basin. Forest birds are hardly seen at all, although the raucous cries of parrots (mostly macaws: arará and maracanã especially) and other songs are heard. Water and shore birds are abundant and visible, especially pretty little swallows that live in holes in river banks and skim rapidly just an inch or so above the water catching insects, and big kingfishers after fishes in the murky water."

Life in Cruzeiro was not dull even though the long delay was maddening. The social life of the town was almost hectic, and there was a continual salvo of firecrackers, on which all the spare (and much non-spare) cash must have been spent. I naively asked what the occasion was and was told that it could be the arrival of an official, the birth of a child, or simple drunkenness. Of the first we had one, of the second several, and of the last innumerable examples. Firecrackers are also *de rigueur* on saints' days, and there is at least one saint for every day of the year.

At 3 P.M. on 26 June, twenty-three days after our arrival in Cruzeiro, we actually started our voyage upriver. Our craft, the *Ajuricaba* (I never learned the source of the name) was a *batelão*, shallow-draft, broad-beam, an open cabin amidship, an outboard motor aft, a forge in the bow serving both for repairs and for cooking. On board were the four paleontologist-geologists, two Brazilian collectors of recent animals who thereafter much of the time went their own way, and four local employees who stayed with us, a cook, a motorman–mechanic, a helmsman, and a man-of-all-work.

My journal of the trip runs to more than two hundred pages and I shall not even abstract them here. We worked hard under trying conditions but saw many marvels in compensation. We collected some fossils, although fewer than we had hoped. We mapped and remapped. We made many geological sections of the river banks. We spent a great deal of our time running aground and getting off again, cutting our way through *pauzadas* (log jams), repairing the motor. When the going was too hard for the batelão we took to a dugout canoe of even shallower draft and much narrower beam (an *ubá*).

On 7 July we reached a place called Taumaturgo, at the confluence of the Rio Amônea with the Alto Juruá. That extreme outpost hardly merits the name of village but it did have several huts, a small trading post, and, remarkably enough, a schoolhouse. We took over the latter, to the well-founded indignation of the schoolmistress, and worked out from this advanced base for some time. *Taumaturgo* means "wizard" or "miracle worker" in Portuguese, but I believe the place was named for an explorer or early settler with that odd surname. The locality has a small place in Brazilian history because there, on 5 November 1904, a day-long battle between Peruvian and Brazilian forces took place. The Brazilians won, and that is why the present border with Peru is considerably farther up the river.

We later worked our arduous way up to the border, where there was a small detachment of Brazilians on one side and Peruvians on the other. On the Brazilian side an Indian war was in full progress, not against the Indians but among them. The savage Amahuacas had recently killed some of the relatively tame Campas, and the most belligerent of the Campas from the river were off to retaliate on the Amahuacas back somewhere in the forest. There on 13 August we made our turnaround and started the long journey back to Cruzeiro and eventually home. As a sample, I kept a minute by minute log of one day's travel (23 August), and I was tempted to copy it here, but it fills three pages with my smallest writing so I give only the final summary:

Total elapsed time, mooring to mooring	9 hours, 1 minute
Moving forward	4 hours, 11 minutes
Stuck, 12 times, total	3 hours, 18 minutes
Other stops, 4 times, total	1 hour, 32 minutes

Distance covered as river flows about 16 kilometers, in air line about 8 kilometers. Average speed per hour, about 1.8 kilometers (about one mile per hour).

Now I must explain our overnight arrangements while traveling. The local men were deathly afraid of the forest and flatly refused to spend a night in it, even at its margin. They slept aboard the batelão, which thus had no room for the hammocks of the rest of us. We climbed up the river bank, often thirty or forty feet and nearly vertical but usually with roots or lianas to help us up. The forest was too dense to camp in, so the workmen would hack out an opening large enough for our hammocks — they would do this much if we stopped before dark.

On 24 August they were making such a clearing when one of the men felled a tree without warning the rest of us. He meant it to fall away from us, but it got away from him and fell right on me. It hit my head, giving me a concussion; my left shoulder, dislocating it; my back, bruising it; and my legs, dislocating my left ankle and shattering my right leg with a compound fracture of tibia and fibula.

I believe that I owe my life to the care of George Whitaker in this emergency. I will not go into some rather horrible details, but just note that I finally got to New York and a hospital eight days after the accident and a series of eventually twelve operations began two days later. It was two years before I could walk again with the aid of a cane. I kept my leg and have been reasonably active, but I am still lame, especially since this condition was exacerbated by an otherwise trivial accident in Indonesia in 1975.

That event changed my life quite radically. The then director of the American Museum removed me from the chairmanship of the department, which was certainly justified while I was disabled. When at last I could get to my office without a wheelchair he explained to me that I would no longer be allowed to travel on any museum business and would only have to report for work in New York. "This is wonderful luck for you," he said. "You don't have to do another bit of work as long as you live. Just clock in every day and then go back home if you like. You'll be paid until you can retire on pension a few years from now."

Although a pension was not immediately due me (I had been on the Museum staff for only thirty-two years) and I had no other job in sight, I resigned rather than accept such a humiliating situation. When this became known I was offered several positions, and I accepted Al Romer's offer of an Alexander Agassiz Professorship at the Museum of Comparative Zoology in Cambridge, Mass., of which Al was then director.

Notes

§ I have now culled epigraphs from two of the small group of British women who have produced what are to my taste the most diverting novels of crime and detection ever written. I also include in that group a woman whose given name (one cannot call it a Christian name) indicates that she was a New Zealander by birth. Like many New Zealanders Ngaio Marsh was more British than most people born in the British Isles. According to the standard Maori dictionary (reference below) the word *ngaio,* difficult or impossible for some English speakers to pronounce, can mean, as a noun, *"Myoporum laetum,* a tree" or "a small grub . . ."; as an adjective, "restless," "deliberate," or "expert"; and as a verb "look carefully at" or "draw figures or patterns." Any of these except the second could apply to Miss Marsh.

§ It is hard to restrain myself from more extended comments on the Portuguese language, which is fascinating in its vagaries. I will only add a note on the stateliness of the language used in polite circles in Brazil. Even friends are normally addressed as "the gentleman" or "the lady," *never* as "you." Extremely close friends or family members can be addressed as "Your grace," abbreviated to *você,* still in the third person. All the innumerable tenses of Portuguese verbs have a second person, but the only time I ever heard it used was when someone addressed a chimpanzee in a zoo. All Portuguese verbs also have an imperative in the books, but I never heard it used. You never tell anyone to do something — not even a bus driver to let you off. You just suggest that if he is in the mood he might be beneficent enough to do it. Even odder is the fact that many Brazilians will go to any lengths to avoid answering "yes" *(sim)* to a question. In a pinch, however, "it is" *(é,* implying that what someone said is so) will do.

§ The obvious casinos, like that at Araxá, had been closed, but gambling continued to be ubiquitous in Brazil, as I am sure it still is. The poor man's gamble was run like the numbers in New York, but it was still called by the odd name *jôgo do bicho,* "game of the critter," because it originated when the director of a zoo used to put a different animal at the entrance every day to excite interest. The bet was on what animal would be there the next day, abandoned when it was realized that no game could be more easily fixed.

§ The translation of the funny joke I heard is São Paulo is:
Father, in a phonograph shop: The kid wants a record. [As some-
times with us, a record is a "disc" *(disco)*.]
Clerk: A record for children?
Father: No. A flying saucer. [In Portuguese they are "flying discs."]

§ We saw Noemí Cattoi later in the United States, but she died
young not long thereafter. She and her successors in the Buenos
Aires museum have been adequate curators, but there has been little
really active and important staff research in vertebrate paleontology
there since Lucas Kraglievich left, long years ago. Such research has,
however, continued in La Plata, Tucumán, and Mar del Plata.

§ Brucellosis has long been epidemic around Mendoza. Minoprio
thinks that Darwin caught it when he was there in the 1830s and
that this was the cause of his subsequent lifelong invalidism. Brucellosis
was never diagnosed in Darwin's lifetime, but his known symptoms
could have been caused by it. I am inclined to agree with Minoprio on
this point, but no one else seems to, preferring such (as it seems to me)
quite unlikely diagnoses as neurosis or poisoning.

§ No one had told the inhabitants of Cruzeiro do Sul that electric
lights can be turned on and off. None had switches and they all
went on when the town generator started and off when it stopped.

§ Piranhas were abundant in the Alto Juruá, but they did not live up
to their bloodthirsty reputation. We often waded among them,
even with our legs bloody from insect bites, and they did not nibble us.
We also ate them — bony but delicious. "Man bites piranha." Perhaps
a separate, peace-loving species?

§ While we were at Taumaturgo we spent much time and effort excavating and preserving what may well have been the largest turtle
ever discovered. It was too large for our batelão, so we left it to be
picked up by a larger government boat when the water was higher.
That was never done. Other fossil reptiles, some excellent, were sent to
Rio for study, but no results have been published. The fossil mammals,
interestingly varied but fragmentary, were to have been studied by me
but that was impossible because of my long disability and then departure from the American Museum, so they too were sent to Rio, where I
suppose they still are.

Besides the collections, which are worthwhile and do still exist, the Juruá expedition made a small but significant geological contribution, the only result yet published. There had long been a theory that during the mid-Cenozoic, more or less, an arm of the sea had extended right across equatorial South America, separating the continent into two parts. Some of the supposed evidence came from the region we examined, and I found that it was incorrect. There was no such transcontinental or inland sea.

§ After some hesitation I have decided to add one grace, or disgrace, note to the circumstances of my departure from the American Museum. Just before I left there the director whose attitude impelled me to leave was removed from the directorship by the trustees and offered a continuing situation there rather like the one he offered me. He accepted.

§ In all fairness I should add that years after I left the American Museum, both in sorrow and in anger, it did begin to pay me a pension. But Anne and I would have starved to death long since if it were our sole source of income.

References

Roe, A. 1953. *The making of a scientist*. New York: Dodd, Mead. This is the book Anne wrote in the winter of 1951-52.

_____. 1956. *The psychology of occupations*. New York: John Wiley and Sons. This is what Anne was working on most of the time while we were in Rio.

Simpson, G. G. 1953a. *Life of the past: an introduction to paleontology*. New Haven and London: Yale University Press.

_____. 1953b. *The major features of evolution*. New York: Columbia University Press.

Simpson, G. G., and C. de Paula Couto. 1957. The mastodonts of Brazil. *Bulletin of the American Museum of Natural History* 112: p. 125-90, plates 1-23. This is the main result of my research with Carlos in Brazil in 1954-55.

Simpson, G. G., J. L. Minoprio, and B. Patterson. 1962. The mammalian fauna of the Divisadero Largo formation, Mendoza, Argentina. *Bulletin of the Museum of Comparative Zoology* 127: 239-93. This is the last and most complete report on the fauna that took me to Mendoza in 1955.

Williams, H. W. 1971. *A dictionary of the Maori language*. 7th ed. Wellington, Australia: A. R. Shearer, government printer. Although only Williams's name appears on the title page, this definitive edition was thoroughly revised and augmented by a committee under the chairmanship of Dr. Pei Te

Huruini Jones, a learned Maori. It is a delightful work, equaled in its general field only by the two volumes of the Hawaiian dictionary by Mary Kawena Pukui and Samuel H. Elbert.

17

Add a Dash of Cacoëthes Scribendi

Tenet insanabile multos scribendi cacoëthes. ["An incurable itch for writing seizes many."] — Decimus Junius Juvenalis.

It has been made clear enough that this account is chronological only overall and not compulsively so in detail. In the last chapter I brought together the two most significant of my several visits to Brazil. In doing so I skipped over some likewise but differently significant matters. To return to them now I must go back briefly even earlier, again to the winter of 1951-52, which we spent in our New Mexican retreat.

One of our visitors there that winter was James Reid, then an officer of the publishing house Harcourt, Brace and Company (later Harcourt, Brace and World, and still later Harcourt Brace Jovanovich). One of the things we discussed was the possibility of my writing a new college textbook of general biology. I was quite hesitant at first, in part because of the magnitude of the undertaking that would be added to my already heavy commitments, which included teaching, a full-time position at the museum, and several research projects. I also hesitated because the whole field of biology seemed too much for one man, especially one who was fundamentally a geologist. Finally I did agree, provided that we could find a good experimental biologist and a good botanist as coauthors. They were found in Colin Pittendrigh, a biologist then at Princeton University, and Lewis Tiffany, a botanist then at Northwestern University.

In 1953 we signed a contract for the book and started work. It was an arduous undertaking. There had to be much rewriting in order to make the work of three authors fairly uniform in style, which was in general

175

my style. A large number of illustrations had to be obtained or, most often, prepared anew, a task in which we all shared but Colin carried the heaviest load. The book entitled *Life* finally appeared in 1957, after my Alto Juruá expedition. I read proofs and supervised preparation of the index as I lay in bed in the hospital or between operations at home trying to recover the use of my shattered leg.

There were exceptions, but up to that time most colleges had separate departments of zoology and botany, and the subject of biology was treated as simply the sum of those two, the study of animals on one hand and plants on the other. Our premise was that there is a general biology that has principles applicable in varying ways and degrees to all living things. In organizing the book we took evolution as the most completely general of such principles, with some physiological and biochemical principles and phenomena secondarily but almost equally general. We also intended to write clearly, without talking down or evading difficulties but minimizing mere jargon. We apparently succeeded in our aims, perhaps almost too much so. The text was widely adopted, and it was copied by a whole succession of competitors, some of whom published what could almost have passed as slight revisions of our book.

Science does not hold still and biology certainly did not in the 1960s, so that in due course our text was clearly in need of a real revision. Lewis Tiffany had retired, and he withdrew from coauthorship. Colin Pittendrigh and I unfortunately could not reach a new understanding, so it was finally agreed that he, too, would withdraw from coauthorship with ample payment for his rights in the work. William S. Beck, a hematologist with unusually broad interests and knowledge in many other aspects of biology, became a very satisfactory new coauthor of the second, extensively revised edition, which after only a little less work than the first edition finally appeared in 1965. The most evident changes were in more extensive treatment of biochemistry and molecular biology in response to a trend among many college teachers of biology. Some other passages which had been written on a rather simple level in the first edition were made more technical and sophisticated. This version, too, was a success, although obviously it could no longer be so original, unusual, or trend-setting as the first edition. Bill Beck was at the Harvard Medical School, as he still is at this writing, and I was then at the Museum of Comparative Zoology, associated with Harvard, so that the collaboration was geographically simpler — only the Charles River separated us.

Another book in the 1950s resulted from a different and delightfully intimate form of collaboration. The idea came to Anne and me sometime in 1953. We remember the incident clearly but are not sure of the date. It was probably a Sunday because she and I were lying late in bed one morning talking about the universe and other things. Psychology is in the main a study of behavior, but up to that time most psychologists took observed behavior as given and paid little or no attention to the fact that it must have originated at some time in the evolutionary history of the species being studied, then usually rats or humans. Such evolutionary concepts as were current in psychology struck me as generally naïve, outdated, or simply wrong. On the other hand, evolutionists were studying mostly morphology, genetics, or to some extent ecology. Some of them did recognize that behavior is also relevant to evolution, but their concepts of behavioral studies in psychology struck Anne as generally naïve, outdated, or simply wrong. We decided to do something about this, got out of bed, and set about doing so.

The idea of holding one or more conferences on behavior and evolution was presented to the American Psychological Association and the Society for the Study of Evolution, the former of which appointed a committee with Anne chairman, the latter, one with me as chairman. The two committees meeting together as a joint steering committee decided to promote an exploratory conference and then, if that indicated sufficient interest and progress, to hold a second conference with the purpose of producing a publishable symposium. The first conference was held at Arden House, Harriman, New York, on 4–8 April 1955. It was clear that there was a basis for a second, symposial conference, and that was held at Princeton Inn, New Jersey, on 30 April – 5 May 1956. Thirty-two people took part in the two conferences, most of them in both conferences although there was some change in personnel.

The symposium, edited and with parts written by Anne and me, was published under the title *Behavior and Evolution* by Yale University Press in 1958. It contains twenty-three chapters by twenty-four authors, under the part titles "The Study of Evolution and Its Record," "The Physical Basis of Behavior," "Categories of Behavior," "The Place of Behavior in the Study of Evolution," and "Evolution and Human Behavior," followed by an epilogue summarizing and integrating the contributions.

That was a seminal work. It strongly influenced the direction of studies both of behavior and of evolution, as attested not only by those who had attended the conferences but also by many of their colleagues and students. As one of many possible examples, in the preface to one of her major works Margaret Mead mentioned the conferences and the resulting symposial book and commented that they had created a necessary focus for her later work and had led to a reexamination of her field methodology.

The lesson of all this is that an effective method for getting really interdisciplinary studies under way is for students of different disciplines to wake up in bed together.

Here I interrupt to point out that I have not been quite fair to myself if I have seemed to imply that an itch for scribbling is the only or the main motive for my extensive writing. Unlike many colleagues and indeed unlike Anne and most of her psychological colleagues, I do not value the spoken word as a means of serious communication. I do not like to lecture, and I do not like to be lectured to, although I have been through more than my share of both. I do not even like oral discussion of anything like a technical topic. I do not readily take in what is said to me, much preferring to have it written when possible. I worry about expressing myself adequately in speech, and still worry but much less about doing so in writing. My *cacoëthes* is not so much *scribendi* as *cognoscendi* or *imbuendi* but is accompanied by a *fastidium loquendi*. (For those whose Latin may be even more rusty than mine, what I think I am saying is that my itch is less for writing than for learning and teaching but it is accompanied by a dislike for talking.)

Someone once asked me how, feeling that way, I got on so well with Anne's colleagues. I explained that they are delightful people, by and large, and that when we entertain a loquacious group of them I just put whiskey on the buffet and flents in my ears (see note).

I have mentioned that my book *The Meaning of Evolution* was not really the text of the Terry Lectures at Yale but that those lectures were a summary of the text of the book. In 1960 I did much the same with the Jesup Lectures at Columbia: I wrote a book, *Principles of Animal Taxonomy*, and summarized it in the lectures. There is nothing strikingly new in that book, but it did sum up a system of principles of systematics underlying the practice of taxonomy and their practical application in the classification of organisms. That system has sometimes been labeled "phylogenetic," but that label has become quite equivocal. It is better called either "evolutionary" or "eclectic." Although now, more than fifteen years later, I would modify it in some respects, I still consider it fundamentally sound and more useful than other systems developed in opposition to it. I will later mention more definitely the controversy that has developed in this field. Here I just say that the rival systems have made some contributions usefully to be incorporated in the eclectic system but that they have not usefully supplanted it. I also think that my now fifteen-year-old book is the best statement of the basic principles of the eclectic system, but with regret that no one has updated it. I do not have time to do so.

I had also already in the 1940s and increasingly into the 1960s been writing essays, as distinct from technical research reports. The essays, too, often corresponded with lectures. Early on I usually lectured from notes, or with slides that were equivalent to notes, and summarized but did not read the written text of a corresponding essay. Later on, perhaps from a failure of nerve reinforced by hearing some of my lectures on tape, I did read more lectures from written texts. Those texts as well as some essays not connected with lectures were published in various periodicals, several as lead articles in the journal *Science*. In the early 1960s I gathered, organized, revised, and supplemented some of them in what is my own favorite among my books, although some readers do not agree with me: *This View of Life*, published by Harcourt, Brace and World in 1964.

In that book seven of the fourteen chapters are revised texts that started out as lectures; five are revised from other previous publications; one is mostly and one entirely new with the book publication. With that selection and revision, the book is not the olio or mishmash usual in a collection of essays but has an organized continuity. It moves from the historical development of modern views on evolution to the nature of science and the place in it of evolutionary science; then to some philosophical problems involving purpose, determinism, and varieties of mysticism; and finally to cosmic aspects of the study of evolution. The title, suggested by the editor, is taken from the opening words of the last sentence in Darwin's *Origin of Species:* "There is grandeur in this view of life. . . ." Incidentally, my friend Stephen Gould, who writes a column for the American Museum's magazine *Natural History*, asked me whether I would mind his using my book title as the title of his column. I pointed out that I had cribbed it from Darwin, and that Darwin could hardly object if he cribbed it too.

My own favorite chapter in this favorite book began life as a lecture at Vassar, where it was enthusiastically received by an overflow audience of delightful young ladies — a reception that doubtless influenced my satisfaction with it. It reviews and compares the views on evolution of Lamarck, Darwin, and Butler. Another, less appealing incident connected with that event is that the train on which I was riding away from Poughkeepsie was stoned or possibly shot at and I was cut by flying glass from the windows. As far as I recall, that was the last train I ever rode in the United States.

The chapter in the book that has caused me the most annoyance is on the nonprevalence of humanoids. In it I expressed the personal, but not unfounded, opinion that it is extremely unlikely that we will ever be in meaningful and useful communication with humanoids elsewhere in the universe. That opinion, which I still hold, has usually been misrepresented, but even those who quote it correctly accuse me of something horrible, such as being opposed to progress or perhaps to motherhood and apple pie.

Another book based largely on previous essays was issued by the same publisher in 1969 under the title *Biology and Man*. It is devoted to the biological sciences and their past, present, and future bearing on the nature and condition of mankind. Five of the ten chapters were revised lecture texts also published in periodicals; two were written as contributions to conferences and previously published as such; three were new essays written for this book. The chapter that I most enjoyed writing is on "Science and the Culture of Our Time." That was prepared for a symposium on "The Impact of Education, Science, and Technology on the South" held at Duke University in 1965. My enjoyment in writing it had some unworthy aspects because it involved making some cracks about beliefs widespread in the South and at the tobacco industry, from which Duke University has so greatly profited. I was unable to attend because of illness, but my paper was read and was also later published in *The South Atlantic Quarterly*.

Another chapter, "Biology and Ethics," was based on an invited address to the American Psychological Association in 1965, later published in *The American Psychologist*. It has had a somewhat mixed reception. Its negative aspects involve condemnation of the naturalistic fallacy ("What is, is right.") and the mystic fallacy ("What God tells us is right."). The positive aspect is that the touchstone for an ethical system is whether it works for betterment in the evolution of the human species. It is not surprising that when I gave the lecture based on this thesis to a large audience a group of nuns rose and stalked out *in medias res*. It is surprising that my old friend Waddington, writing his own book on ethics, completely misunderstood what I had written — he must have misunderstood it because I cannot believe that he would purposely so completely misrepresent it.

Some people like this book better than *This View of Life,* and it has been almost as widely translated, but I still prefer the earlier book of these two. Some people do not like either one, and I wish both were better, but then everything I have done was just the best I could do at the time and I wish everything could have been better.

Notes

§ The authors and editors of *Behavior and Evolution* waived their rights, and royalties were paid to the two sponsoring societies, the American Psychological Association and the Society for the Study of Evolution. The book finally did become outdated, largely because of studies that it had itself stimulated, and eventually it went out of print, but substantial royalities were paid to the societies for almost twenty years.

§ Margaret Mead and I were associated at the American Museum for many years, although I never entered her aerie in a remote corner of that building. For a time she was one of a small coterie of women anthropologists who took what I considered a mistaken and invasive interest in my marital problems, but Margaret and I later became and I trust still are on a reasonably friendly footing.

§ I don't find *flents* defined in a dictionary. This is a possibly proprietary name for a kind of ear plug with which I became acquainted during long stays in noisy hospitals.

§ One of the chapters in *Biology and Man* — I think it advisable not to specify which one — was written for a conference that was in general so inane that I manufactured an excuse for leaving in the middle of it. Without my knowledge or permission a hopelessly garbled version of my paper for the conference was published. That reminds me that a Russian translation of one of my books — the only Russian translation of any of them, as far as I know — was also published without my knowledge or permission, and of course without royalities. I found out about its existence only by accident. I did know that a Spanish translation of another book had been made because I painstakingly corrected the translation in manuscript, but I did not find out that it had in fact been published, without much attention to my corrections, until inquiries were made several years later by the United States Information Agency. But almost any author could match or top further horror stories about adventures with editors, publishers, and translators.

References

Dobzhansky, Th. 1967. *The biology of ultimate concern.* New York: New American Library. I put this in as an extra because it bears on topics in *This View of Life* and later in *Biology and Man*; I do not entirely share Dobzhansky's views on philosophy and theology, but I highly respect them.

Mead, M. 1964. *Continuities in cultural evolution.* New Haven and London: Yale University Press. This is the book that was inspired in part by Margaret's valuable contribution to our second conference on behavior and evolution.

Roe, A., and G. G. Simpson, eds. 1958. *Behavior and evolution.* New Haven and London: Yale University Press.

Simpson, G. G. 1964. *This view of life: the world of an evolutionist.* New York: Harcourt, Brace and World.

———. 1969. *Biology and man.* New York: Harcourt, Brace and World.

Simpson, G. G., C. S. Pittendrigh, and L. H. Tiffany. 1957. *Life.* New York: Harcourt, Brace and Company.

Simpson, G. G., and W. S. Beck. 1965. *Life.* New York: Chicago, Burlingame: Harcourt, Brace and World.

Waddington, C. H. 1960. *The ethical animal.* London: George Allen and Unwin. This is a valuable evolutionary approach to ethics despite the facts that brief reference to my own views completely misrepresents them and that I find its authoritarianism unacceptable.

18

Fresh Woods

Fresh woods and pastures new. — John Milton

In 1959 Anne and I moved from New York to Cambridge, Massachusetts, where she soon joined the faculty of Harvard University and I became an Alexander Agassiz Professor in the Museum of Comparative Zoology at Harvard College. Once when I used the latter correct designation an old grad condescendingly told me that the M.C.Z. (as it will hereafter be called) is not *at* Harvard *College* but *of* Harvard *University*. In fact Harvard was a college and not a university in 1859, when the M.C.Z. was founded, and the M.C.Z. was chartered by the state as a separate institution at but not of Harvard College. The museum and the university are indeed closely interrelated, but the relationship is symbiotic and not filial.

The M.C.Z. was founded by Louis Agassiz, a distinguished French-Swiss zoologist, paleontologist, and geologist who moved to the United States in 1846 and became a professor at Harvard. His son by his first wife (née Cécile Braun) was Alexander Agassiz, who became an accomplished oceanographer and also, through investment and management in Michigan copper mines, a wealthy man. His will left large trust funds first for members of the family and eventually for the M.C.Z. When I went to the M.C.Z. just a hundred years after its founding the capital of these funds was beginning to be turned over to the museum and to be used, as the will had directed, for professorships in the M.C.Z., not in Harvard. Because the Alexander Agassiz professorships are for set terms, without tenure although renewable, Harvard also made me professor of biology and professor of geology, with tenure but without pay unless, as did not quite happen, M.C.Z. ceased to pay me while I still had tenure at Harvard. This explains my previously mysterious comment that for a time I held three professorships simultaneously.

This move was radical for us in the literal meaning of the word: it affected the roots of our lives. Nevertheless we soon settled in quite happily among new associates and friends, in a new scene, and in new ways of life. It is true that eventually some unforeseen and, as it seemed to us, unnecessary drawbacks appeared. As I look back on those years my feelings about the M.C.Z., but not Harvard, are ambivalent. That is true to such an extent that I have found it hard to start this chapter — this is my fourth draft of the first few pages. Still my experiences at and with the M.C.Z. were much more positive than negative, and at this point it would be petty to accentuate the negative. Harvard, as such, was good to us. Before long Anne was made a full professor in the Graduate School of Education, where she founded and ran a Center for the Study of Careers. She was the ninth woman professor in the well over three hundred years of Harvard's history (women have done better there more recently). The president, then Nathan Pusey, was fond of pointing out that we were the only Harvard professors ever to be married to each other — and as far as I know, we may still have that distinction.

In the M.C.Z. I was given a large but rather Spartan office on the second floor, convenient to the elevator, the men's toilet (women had to go to a different floor), the library, and eventually the smoking room — for years smoking was prohibited anywhere in the building, with the result that staff spent much time on the outdoor steps or surreptitiously smoking in the toilets. I shocked the director, Al Romer at that time, by having my dingy office walls repainted in bright, cheerful color. Although frequently informal and even ebullient in company and before audiences, Al was rigidly conservative in many more personal ways. Although he detested beards and forced a student to shave his, he tolerated mine on the grounds that beards over twenty years old could be sanctioned. I never gathered how one could have a twenty-year-old beard without ever starting one.

The collections and field notes made while I was at the American Museum remained with that museum and I necessarily abandoned previous research projects except the partly published monograph on the earliest Patagonian mammalian faunas. The specimens not yet adequately published were sent on loan to the M.C.Z., and I gave first research priority to the completion of that study. Still it was eight years before the final volume came off the press. In the meantime work was done on some of the books mentioned in the previous chapter, especially *Principles of Animal Taxonomy,* written early in my stay at Cambridge. I also undertook some new, relatively short research projects. The most interesting were perhaps one on the historical zoogeography of Australian mammals and one on species density in North American recent mammals. The former continued my lifelong interest in marsupials and brought in an aspect to which I have recently returned. The latter was a new idea at the time, since followed by several similar studies by others. I plotted the numbers of species in quadrates of equal areas covering the whole continent and then mapped this by contour lines following equal numbers, a sort of isograms or isopleths.

Alexander Agassiz's will specified that professors employed in the M.C.Z. on his endowment should not be required to teach, but there was strong pressure to do so. From 1960 into 1967 I did teach one graduate course on evolution jointly with Ernst Mayr, who became director of the M.C.Z. when Al Romer retired. I also had numerous other contacts with students both in biology and in geology. On the whole this was more satisfactory than had usually been true at Columbia. The Harvard – M.C.Z. community in Cambridge was more close-knit than the Columbia – American Museum community could be in a city like New York, and we soon had as many pleasant social contacts as we desired.

We had not been exactly stay-at-homes before moving to Cambridge; in fact, I had at least touched on every continent except Antarctica (that was to come later). But we now started even more extensive travels. Since 1959 there have been few years without at least one long journey and none without short trips. Usually business and pleasure have been combined, a conjunction that arises almost automatically in my profession and to one of my temperament. Most of our travels while I was on the M.C.Z. staff were in summer, when I was not required to be in residence but was expected to be professionally occupied.

Our unusually varied program for the summer of 1960 started with attendance at the tercentenary of the Royal Society, of which I have been a foreign member since 1958. That was quite grand, with a garden party at Buckingham Palace, addresses by the queen of England and the king of Sweden in the Albert Hall with all the R.S.'s berobed, a tea for R.S. wives with the Queen Mother in St. James's Palace, a tea for all in the Houses of Parliament, a conversazione graced by the duke of Edinburgh in Burlington House (where the R.S. was then quartered; it has moved since then), a lunch at the elegant country estate of Sir Thomas Merton, a reception by the lord mayor in the Guild Hall, and finally an enormous dinner addressed by Macmillan, then prime minister — among numerous other events and personages. Not the most impressive but the most amusing event occurred at the Guild Hall do, when the honorary gentlemen pikemen came clumping in, out of step and with their pikes at odd angles, clad in armor contemporaneous with the founding of the Royal Society in 1660. I have done a bit of name dropping and could have done more, because London was crowded with regal and noble personages from all over the world. I'll drop one more name, or a pair, for one day when I was on Fleet Street along came the Horse Guards, gorgeously caparisoned, followed by an open horse-drawn carriage in which were the king of Siam (or Thailand) and his extraordinarily beautiful queen.

It was also on this occasion that we visited Down House, and I did get in a bit of paleontology at University College, but not at the British Museum (Natural History). There I was most pleasantly greeted by the director, then Morrison-Scott, but to my amazement and for reasons never explained I not only was not greeted by the paleontologists but was refused entrance to their department — something that has never happened to me on any other occasion either there or anywhere in the world. I must add that my old friend Hopwood had retired, that the paleontologists then at the B.M. (N.H.) are no longer there, and that those now there had not then joined that staff.

From England we flew to Spain to see fossils and fossil localities and to meet a group of Spanish paleontologists, with one of whom, Miguel Crusafont Pairó, I had long corresponded but whom I had never met. In Barcelona we managed to rent a very decrepit Seat, which at that time was difficult enough because car rental was still almost unknown in Spain. It was made even more difficult because the young man handling the transaction twice lost the car keys, parked the car on a side street and forgot where it was, then parked it in a no-parking zone, drove it the wrong way on a one-way street, collided nonfatally with another car, left the lights on and discharged the batteries, and committed so many other idiocies that we nicknamed him *El Despistado,* which in Spanish slang means more or less "off the rails." A Seat is a relatively cheap and positively inferior Spanish version of a Fiat. There is, or was, a saying that those who go anywhere in Spain are either *Seatones* or *Pietones,* meaning that you either travel in a poor car or you walk (I don't guarantee the spellings; neither word is in my dictionaries).

We had been met by Sr. Jaime Truyols Santonja and Fr. Emiliano de Aguirre, both of whom had been working with Crusafont at Sabadell, a smaller city near Barcelona, and by Sra. de Truyols. I could no longer drive and none of our Spanish friends at that time had ever learned to, so Anne had to drive us over a difficult but fascinating route to where we were to meet Crusafont at a distant place in the Pyrenees. We started out along the main highway toward Madrid, which was paved, but barely two cars wide and almost without shoulders. There were a few filling stations on that highway, but none after we left it. Thereafter gasoline could be obtained, if at all, in village garages and if you also wanted oil, air, or a repair of any sort you had to get to a larger town with several garages and go from one to another.

We first went right across the Vallés-Penedés Basin, rich in fossil mammals — the European geological age and stage Vallesian is named from it. After that we were in hilly country, almost every hill crowned by an abandoned and ruined castle, relics of the time when this was the frontier battle zone of Christians against Muslims. There are hundreds of castles in Spain and, except as scenery, they are indeed worthless. Hillcrest villages dating from the same period are, however, still inhabited. Hence we turned north into the pre-Pyrenees and up a very narrow, twisting road high along a cliff-side with a burro likely to appear around any blind turn. Then through a tunnel some five kilometers (about three miles) long under the jagged crest of the Malditos — "Accursed Mountains" — and out into the Valle de Arán, a longitudinal valley in the highest Pyrenees.

That rich, green, incredibly beautiful valley has been heavily populated since the Middle Ages. The entire population lives in thirty-two tiny hill villages in a space of only a dozen miles, so that some villages are only a few hundred yards apart and yet are perfectly distinct. In the village of Salardú we stopped at a small, pleasant resort inn where, after twenty years of pen-palship, I finally met Miguel Crusafont and his wife. There also were two French paleontologists, Bergounioux and Croizet, the former a Franciscan friar and the latter an abbé. Politically Spanish, this valley is topographically French — its river is the head of the French river Garonne. The two clerics had come up from Toulouse to meet us. The Crusafonts had recently bought one of the old, high massive, fieldstone-walled and slate-roofed farm houses and were starting to have the interior remodeled and partly modernized.

After a tremendous lunch we spent a long rainy afternoon discussing evolution and theology in four languages: Spanish, Catalán, English, and French, but mostly French because the clerics from France spoke no other language although they had spent much time in Portugal and Spain. The Crusafonts are bilingual and speak Spanish fluently enough, but their native language is Catalán and they speak it between themselves. It is the language of Barcelona and the rest of Catalonia and is quite unlike Spanish, although almost identical with the Provençal spoken just across the border in southern France. Truyols and Aguirre are Spanish (Galician), not Catalán. We all had a smattering, at least, of French.

All but Anne and I were devout Roman Catholics but all the men were also evolutionists (with some reservations), a stand then decidedly rare among Spanish Catholics. Even Crusafont was taken aback by Fr. Bergounioux's more worldly views, especially his rejection of Teilhard's mysticism, and Aguirre later admitted that he was somewhat surprised at the evident deference shown me by the French men of God.

All of the men, including the priests, were jocular and constantly amusing with volleys of puns and jokes. Fr. Bergounioux called himself "Le père rigolo" and Fr. Aguirre told anticlerical and dangerously political jokes. I will give one of each as samples:

1. A couple on the way to their wedding was killed in an accident and went to heaven. They asked St. Peter. "We were just about to be married, so can't we be married here to make us happy?" St. Peter replied. "I'm sorry, but it is impossible to be married in heaven. There are no priests here."

2. Franco was flying around and gave the pilot a lot of pesetas to drop to the people below. He asked, "Are they happy now?" The pilot replied, "No. They want more." So Franco had the pilot throw out five peseta pieces, and so on until thousand peseta notes were being dropped. "Surely they're happy now?" "No, they won't be happy until I throw you out."

From Salardú we drove back to Sabadell and spent some time between there and Barcelona. With Crusafont I went over his fine collection of fossil mammals, visited a number of fossil localities, and made a detailed study of the Vallés Penedés. We also did some sightseeing, and I gave a lecture to a large and courteously attentive audience in Sabadell. I could not speak their language, Catalán, and the Argentinian-American accent of my Spanish must have puzzled them a little, but they did follow it, and they conducted the meeting in Castilian, which we found easy to understand. (On another occasion in Sabadell I attended a meeting conducted entirely in Catalán and could follow little of the proceedings.)

Crusafont is an interesting and admirable man. He and his wife are both pharmacists and had two pharmacies in Sabadell. He contracted an intense passion for fossils and became a self-taught and well-taught paleontologist. For years he devoted as much as possible of his time and resources to collecting ancient mammals. In 1960 the collection was inadequately housed in an old building, but since then the municipality has erected an excellent, modern museum, which Crusafont now directs. He has also become a professor at the University of Barcelona and received other recognitions for his remarkable achievements. Like our other Spanish acquaintances, he is also *chistoso* (humorous, facetious), which reminds me that as a pharmacist he modified the standard Spanish toast *Salud y pesetas* ("health and money") to *Salud pero no demasiado* ("health but not too much of it"). He also hispanicized our toast "More and better fossils" to *Más y mejores fósiles*.

From Sabadell we drove to central and southern Spain, with Emiliano Aguirre accompanying us as far as Madrid. Madrid has not become one of our favorite cities, but we passed several pleasant days there. Considerable time was of course spent with paleontologists of the Universidad Central (the University of Madrid), and apart from that we most enjoyed the Prado with its wonderful Goyas. I also especially admired the Hieronymus Bosches, but Anne does not like them. *Las Meninas* by Velásquez is also in the Prado and we found that we could see them only through the eyes of Picasso — while we were in London there was a big retrospective show of Picassos at the Tate and it included numerous variations based on *Las Meninas*. We also went to Toledo for a day. The whole town is charming in a touristy sort of way, but we especially wanted to see the *Entierro del Conde Orgaz* by El Greco, which is not in a museum but in a small church above the tomb of the count. We both consider it probably the finest picture ever painted.

From Madrid we went on to Córdoba, where we were shocked at the desecration of what had been a splendid mosque by the building of a trumpery, baroque (or in more Spanish terms *plateresque* and *churrigueresque*) Christian cathedral right in the middle of it. On to a couple of days in Seville, a fascinating city where we enjoyed a visit to the Archivo de Indias, repository of all documents pertaining to the discovery, exploration, and colonies of Spanish America. It is a working institution and does not encourage tourists but does have a few manuscripts out in display cases, including a touching letter to a boy "from your father who loves you" — Christopher Columbus.

Thence to Granada, where at that time Aguirre was both studying and teaching at the university. He had left us at Madrid and gone back to Sabadell, but his assistant, a charming young woman, greeted us and showed us about the small but busy geological department. With her we also went to the archaeological museum, presided over by a *directora* who impressed me in several ways, among them by being able to read Arabic in kufic characters, which few people whose native language is Arabic can do. (I can cope with naskhi characters, but kufic is Greek to me; so is most Greek, for that matter.)

On this visit to Granada we soon went to the Alhambra several times and found it all our fancy had painted it and more. We had not even been aware that it is not one building but a whole fortified settlement. For descriptions, see any good guidebook to Spain. After those first visits we went back twice with the directora, who was well known and a sort of pet of all the guards and officials. Now we had VIP treatment, were allowed into rooms not open to the public and permitted to stay when the Alhambra as a whole was closed to tourists and we could enjoy the place without mobs of other visitors. Going over after hours to the Generalife, a summer park and palace near but outside the Alhambra, a group of Franciscan friars from Córdoba, known to the directora, attached themselves to us. The gatekeeper balked at letting them in, but the directora introduced me as the *great* Professor Seemsone and the friars as my students. They were let in, but we all spoiled the effect somewhat by giggling.

Our next considerable stop was in Valencia, where the attraction was an extraordinary large collection of Pampean (Pleistocene or Ice Age) fossil mammals from Argentina, the most considerable such collection outside of South America. The collection is in this oddly out of the way place because it was presented to the city by an engineer, J. Rodrigo Botet, who was born in Valencia, worked for some years in Argentina, and collected fossils as a hobby. The specimens are installed in a medieval building, originally an almudín or granary, that is itself a museum piece. The collection has been well cared for by a succession of directors of the museum, the Boscás, father and son, Beltrán (who was director at the time of our visit), and Martel. A complete illustrated catalogue by Martel and Aguirre was published in 1964. Nevertheless it must be said that a really thorough study of this exceptional collection has yet to be made.

We got back to Barcelona on 17 August, went to Sabadell for an affectionate farewell to the Crusafonts and Aguirre, and on the nine-

teenth went by train to Les Eyzies, a village in the Dordogne region of
France. The train trip required four changes, all difficult, and we
arrived late at night in pitch dark at a station apparently in a wilderness.
We did finally locate our small hotel, the Cro Magnon, and were given
a much desired cold collation by the friendly proprietors. We were
reminded of the New Englander's wisecrack, "You can't get there from
here." We came close to being convinced that you can't get to Les
Eyzies from Barcelona. Speaking of wisecracks reminds me of the part-
ing one from Fr. Aguirre:

Small child, on catching sight of a priest's trousers under his robe: Papa!
There's a man underneath that padre!

The Cro Magnon Hotel was on the very spot where Cro Magnon
man, one of the earliest known representatives of *Homo sapiens,* was
found. The bathroom (there was only one) had the bedrock of the prehistoric shelter exposed as part of its roof and a wall, which made Anne
uncomfortable. A couple of hundred yards from there a group of students working with Professor Movius, whom we knew at Harvard,
were excavating a similar shelter. It was in such open nooks under rock
overhangs, what the French call *abris,* and not in what anyone would
properly call a cave, that many of the Paleolithic (Old Stone Age) so-
called cave men lived. They did enter some caves but apparently for
ceremonial reasons and not to live in them.

While staying in Les Eyzies we hired the only taxi in town and went
up the lush and intensively cultivated valley of the Vézère past La
Madeleine and Le Moustier, for which the Magdalenian and Mousterian
subdivisions of the Old Stone Age are named, to the Grotte de Lascaux.
This cave, discovered, or rediscovered, by accident in 1940, has the best
preserved and most beautiful of all known Paleolithic art, a really stunning accomplishment by such ancient artists. We were fortunate to see
those paintings, as it was later found that the humidity resulting from
admitting large groups was causing them to deteriorate and the cave
was closed to the public. We also took a merely scenic tour by our
trusty taxi along the Dordogne Valley and past its many châteaux. We
were interested to hear that Josephine Baker then lived in one of them
with a number of adopted children. She was an American black woman
who had danced nude in Paris when she was younger, and I had seen
her do so when I, too, was younger. After 1960 she had some financial
problems and she is now dead. She was a remarkable and, I feel, on the
whole admirable woman.

On 26 August we took a night train to Paris, also very uncomfortable but at least with no changes. Anne unfortunately had to return then to the United States, but I stayed on for some ten days in Paris. I had made a dozen visits to that city over the past thirty-seven years and was no longer entranced with it, but I worked diligently in the paleontological gallery at the Jardin des Plantes and did greatly enjoy the hospitality of some colleagues, especially Don Russell, an American who was settling in permanently there (he is still there in 1976), and Robert Hofstetter, who is utterly French in spite of the fact that like many Frenchmen from eastern France he has a German name. I also took a one-day excursion with those two to the region of Rheims, known to oenologists for its champagne but to paleontologists for its early (Paleocene and Eocene) mammals, on which Don was working.

On 6 September I flew to Brussels to take part in a colloquium on early and primitive mammals arranged by G. Vandebroek of the University of Louvain. Two of my M.C.Z. colleagues, Romer and Patterson, were there. So were Crusafont and some eleven other paleontologists, mostly old friends by now. Then I went to Oxford for a seminar on comparative anatomy and vertebrate paleontology, attended by virtually all British vertebrate paleontologists and numerous others, and finally back to Cambridge (U.S.A.) on 16 September.

Here I will again become nonchronological and skip to our second and, up to the present, last visit to Spain. That was in 1969, and there is much relating to the period between 1960 and 1969 to which I will return. The primary occasion for this return to Spain was an invitation from the University of Salamanca to take part in the First International Symposium on Animal Phylogeny *(I Simposio Internacional de Zoofilogenía)*, to take place in that venerable institution in October. Our interest in this was reenforced by an invitation from the Crusafonts to go with them to the symposium and to tour much of northwestern Spain still unknown to us. In response to that invitation we went first to Sabadell, via Barcelona, and set out from there on quite an indirect route to Salamanca.

This time there was no car rental to manage and no driving for us to do because Señorita (now Doctora) Juana María Golpe Posse, a student and colleague of Crusafont's, provided a car and did the driving. Miss Golpe is a delightful, highly dynamic Galician *(Gallega)* and we greatly enjoyed both her company and her chauffeuring. We went first to the Valle de Arán, which we had briefly admired in 1960. The remodeling of the Crusafont's farmhouse had been completed, and we stayed with them there. With some relaxing, enjoying the scenery, and visiting some of the charming villages, I spent most of my time putting my paper for the symposium into Spanish, with Crusafont and Golpe helping to put my still somewhat Patagonian Spanish into more fully idiomatic Castilian.

The University of Salamanca was founded in 1230 A.D. and is the oldest in Spain. At its height, in the sixteenth century, it was also the most progressive and, with about ten thousand students, the largest. Thereafter it slowly declined, almost to extinction. In spite of a reorganization in 1857, by 1900 it was the smallest and among the most backward of the Spanish universities. Since then it has made much progress and the symposium was an example of increasing interest in science and, it was hoped, a stimulus for further development. We were lodged in a comfortable motel a few miles out of town, where we were pleased to find Emiliano Aguirre, Bernhard Rensch down from Germany, and other friends old and new. One of the new friends, who has continued as such ever since, was Francisco Ayala, Spanish in origin but American by adoption.

The sessions were held in one of the attractive but in truth not entirely comfortable old buildings of the university. The papers presented were something of a mixed lot, as seems to be inevitable at such symposia, but included much of interest. They were given in four different languages, mostly the native languages of the speakers, although the two Danes spoke in English, the one Rumanian spoke in French, and I spoke in Spanish. (In the published proceedings the editors insisted on printing the English version of my paper, not the Spanish version that three of us had so carefully produced.) There was considerable pomp and ceremony and much drinking of good wine together, as befits a symposium (Greek *symposion,* "drinking together, a drinking party"). By way of relaxation, there was also an excursion to Ciudad Rodrigo, an old and still rather medieval town near the border with Portugal. Here there was a smith making beautifully decorated steelyards. These, *romanas* in Spanish, are still the usual means of measuring weights in that region, but they impressed us as works of art and one now hangs in our living room, where it baffles guests at first sight.

On our way to Salamanca we had gone through Ávila, a famous old walled city, former home of Saint Teresa, and one of the major Catholic religious centers. In further touring we stopped in Oviedo and visited the Truyols, who had moved there from Sabadell. In the bustling and historic city of Coruña, now capital of Galicia, we were most hospitably received by Miss Golpe's parents. We also went to Santiago de Compostela, which is the most frequented place of pilgrimage in Spain and ranks with Rome, Jerusalem, and Mecca as among the principal ones in the world. There seem to be places called Santiago in most Spanish-speaking countries, but this one is supreme because the Congregation of Rites declared (more than 1,800 years after his death) that the body of the apostle James or St. James the Greater (Santiago in Spanish) is indeed preserved in the cathedral of this city. We duly visited the cathedral, which has an honest romanesque basis but is overlain with churrigueresque nonsense or worse. Anyone who has read this far knows that I am not pious in any orthodox sense and will not be too surprised that my clearest memory of Santiago de Compostela is that we there had a delicious lunch that included the first barnacles (goose barnacles or lepadids, not rock barnacles or balanids, which are inedible) that Anne and I had ever eaten.

So home again with satisfaction at now having had a glimpse, at least, of all the major regions of Spain and fair knowledge of its geology and the history of its mammals. Home now not to Cambridge but to Tucson for reasons that will be filled in later.

Notes

§ The epigraph of this chapter is usually misquoted as "Fresh fields . . ."; Milton wrote "Fresh woods. . . ."

§ When Louis Agassiz went to the United States he necessarily left behind his ill European wife. She died in 1848. In 1850 Agassiz, now permanently settled in Cambridge and active as a professor and public lecturer, married Elizabeth Cabot Cary, a proper Bostonian. She founded Radcliffe College and was its first president. That college has no faculty and its students have always been taught by Harvard professors, for years in segregated classes but now in regular Harvard classes leading to Harvard degrees.

§ In America the name Agassiz is pronounced in somewhat the French way but without a French accent: *ag'*-uh-see.

§ Apropos of Al Romer's retirement and Ernst Mayr's appointment as director of the M.C.Z., I think that it is now no breach of confidence to reveal that the appointing body, anomalously called the "faculty" at the M.C.Z. but equivalent to the trustees, offered me the directorship. I declined and recommended Ernst instead.

§ Tercentenary is one of quite a few words that look alike but do not sound alike in American and British English. In American it is ter-*sen'*-ten-ary, in British ter-sen-*teen'*-ary. The Britons do have more tercentenaries than we do, although Harvard's was already past at this time.

§ The peculiarity of the Catalán language can be illustrated by a single word, *puig*, which means "hill" and is much used in that hilly country. The orthography of Catalán is peculiar and *puig* is pronounced "pooch." Our companions were delighted when I told them that *puig* (as pronounced) means "a dog of doubtful ancestry" in American English. A more complex bilingual pun that enchanted them is that "Look out! You'll fall!" in Catalán sounds like *I see cows* in English.

§ While going over field localities Crusafont and I noted that rich deposits of fossil mammals seemed often to coincide with regions with good local wine. This correlation seemed worthy of careful study and we planned to seek a large grant for research on it, but somehow the plan did not work out.

§ The Alhambra is the usual expression but it pains me because it is redundant: the *al* in *Alhambra* means "the." The Arabic name was *al Hamra* (approximately), "the red [one, or house]." This reminds me that dry watercourses in southern Spain are called *ramblas* and that some streets in Barcelona are also called *ramblas*. A guide assured a party of English tourists that this is because people ramble along them. In fact they are so called because they were built over what had been sandy watercourses and an Arabic word for "sand" is *rammla* (more or less — one can't write proper Arabic with Latin letters).

§ *Generalife* is another scrambling of an Arabic term: *Jannatal-Arif*, "raised garden."

§ The tiny, nominally independent republic of Andorra lies eastward in another physiographically anomalous valley of the Pyrenees such as the Valle de Arán, but the latter is politically entirely Spanish. The Basque country, Spanish by force but like Catalonia separatist at heart, lies west of the Valle de Arán.

§ I do not know of any second symposium that has yet followed the first on animal phylogeny.

§ Coruña (called Corunna by the British) dates from Phoenician times and like most Spanish cities, or indeed most European cities, it has repeatedly been fought over. To English-speaking peoples it is best known by an eloquent dirge on the death there of Sir John Moore in 1809 during the Peninsular War.

References

Alvarado, R., E. Gader, and A. de Haro, eds. 1971. *Actas del I simposio internacional de zoofilogenía.* Universidad de Salamanca. Texts of papers presented in 1969. My contribution, "Status and Problems of Vertebrate Phylogeny," is on pages 353-68.

Lurie, E. 1960. *Louis Agassiz: a life in science.* Chicago: University of Chicago Press. The best of the several biographies of Louis Agassiz.

Simpson, G. G. 1961a. La evolución de los mamíferos Sudamericanos. *Estudios Geológicos,* 17: p. 49-58. This is the text of the lecture I gave at Sabadell on 3 August 1960.

————. 1961b. Historical zoogeography of Australian mammals. *Evolution* 15: pp. 431-446.

————. 1964. Species density of North American recent mammals. *Systematic Zoology* 13: pp. 57-73.

Tharp, L. H. 1959. *Adventurous alliance: the story of the Agassiz Family of Boston.* Boston and Toronto: Little, Brown. A collective biography of the family, but with most emphasis on Elizabeth Carey Agassiz, Louis Agassiz's second wife.

Vandebroek, G., ed. 1961. *International colloquium on the evolution of lower and non-specialized mammals.* Brussels: Vlaamse Akademie Voos Wetenschappen etc. Texts of papers at the 1960 colloquium. My contribution, "Evolution of Mesozoic Mammals," is on pages 57-95.

19

Ex Africa Semper Aliquid Novi

I speak of Africa and golden joys — Shakespeare

Pliny's Latin version of an old Greek saying is so familiar that I use it as a title without translation. The epigraph is no more original, but it is less familiar. Both are so appropriate as to be practically obligatory for this chapter, in which I intend to discuss our several postwar visits to Africa. The first and most important one fits in chronologically after our first visit to Spain, for it occurred in the next year, 1961.

On that first visit since the war, and my first to sub-Saharan Africa, we flew to London and stopped over there for a couple of days. We visited there with Julian and Juliette Huxley and with Ethel John Lindgren, a distant cousin of Anne's and a former explorer of central Asia, now married to a Lapp reindeer breeder named Utsi. Thence overnight to Entebbe, Uganda, where we had breakfast at the airport on the shore of Lake Victoria, and on to Nairobi. We were met by two of the zoologists from the Coryndon Museum and installed in a pleasant, small hotel, Ainsworth, at easy walking distance from the museum. The Coryndon Museum, then named after a former British governor of Kenya and now renamed as the National Museum of Kenya, was in the outskirts of the city. It then contained not only exhibits and large study collections of recent animals but also the constantly growing collection of fossils being made by the Leakeys, for whom this museum was a city base and research center. It was primarily on that account that we were visiting Africa.

It was a few days before we met the Leakeys because Louis was delayed in Paris and Mary was caring for a sick son, but we started forthwith to observe both the living and the ancient faunas. In that connection it is interesting that there was a Ghandi Hall in the museum, presented by East Indians, who specified that nothing killed for the purpose should ever be exhibited in that hall. Consequently, in addition to mineralogical and ethnological exhibits the fossils were exhibited there — dead animals, to be sure, but clearly not killed by their collectors. That reminds me of the story, true, I believe, of a British collector who came to Kenya and was eaten by crocodiles. Someone asked whether he had collected reptiles and the reply was "No. On the contrary."

On our second day in Kenya we hired a car and started out early enough to reach the Nairobi National Park just at sunrise. We came back exalted with delight. My journal account for the day starts, "I find it hard to put things in order or to express one of the greatest experiences we have ever had. The Age of Mammals, living, breathing, and kicking up its heels. Hundreds and hundreds of animals everywhere, graceful, awkward, comical, sinister, beautiful, ugly — to match any adjective." Since then we have seen many such sights, always with joy, but the impact of the first acquaintance has not faded. I decided even then, and have never changed my mind, that a male impala is the second most beautiful creature in the world (the first is of course a female human). Male impalas are more beautiful than female because they have graceful horns and the females are hornless. Zebras were at once characterized as "clean, tidy, dainty, neat, sweet." Wildebeests (gnus) "frisky." Warthogs "ugly as sin but endearing in a comical sort of way." Giraffes "unreal." Our luckiest sighting was a pair of leopards returning from the night's hunt. Some residents of Nairobi who had themselves never seen a leopard supposed that I had mistaken cheetahs for leopards, but I did know the difference and I had a photograph to prove it.

The next day was equally fascinating in a different way. Shirley Coryndon, who worked at the museum named for her father-in-law, drove us to Lake Magadi in the great Rift Valley. Environments different from those in the Nairobi Park added a number of new animals to our sighting list, notably gerenuks, slender antelopes that browse high on trees by standing on stiltlike back legs, and hyraxes, the "feeble folk" of the Bible although even from that authority I reject the adjective. In the rift bottom we also had a first view of the seminomadic Masai, quite different from the sedentary Kikuyu around Nairobi. The water of Magadi is saturated with salt and soda and the thick crust of soda is calcined there and shipped by rail to the coast. It is used mainly for making glass, but there had been a recent order from Yemen for 300 tons of unrefined soda to be mixed with snuff!

Driving back, Mrs. Coryndon told us that some of the giraffes had become, as she put it, "naughty" lately. They had learned that they could stop an automobile by rearing up and stamping on it with their front feet. A driver had recently had his car totaled in that way and the insurance company disbelieved the report "destroyed by giraffe" but finally did pay up. The driver purchased a new car and drove it down this road, where it, too, was promptly wrecked by the same giraffe. This made us a bit nervous when there were giraffes in our way, but they let us pass. That evening at the hotel a pretty, delicate-looking young English girl introduced herself and told us she had just spent seven months watching chimpanzees. That was Jane Goodall, who has since become famous for her studies of those animals.

My visit to Africa was supported by a Guggenheim Fellowship and was not just for enthusiastic animal viewing, although that legitimately was included, but for study of fossil mammals and of the strata in which they occur. I worked for some time in the Coryndon Museum, going over the whole of the fossil collections but especially the primates, members of the order to which we belong, and among them especially the lorisiforms, primates such as the bush-babies and pottos that have remained at a primitive evolutionary level. I eventually published a report on all the known African fossils of that group, but I mention this with some hesitation. Most of that study still seems sound and useful, but it includes one howling mistake, possibly excusable but the sort of thing that gives a conscientious scientist nightmares. I try, sometimes in vain, to reassure myself with the thought, which I do believe to be correct, that those who never make mistakes never make much of anything.

Louis Leakey was ill for a while when he got back to Nairobi, but on 10 June we were able to set off with him in a Land Rover on the first of our visits to fossil localities. These visits were primarily for my own information and education, but they also had a somewhat ulterior motive that I never made fully explicit to Louis. He was a strongly positive character, something of a showman, with sometimes rather exaggerated enthusiasm that could resemble sensationalism. This led some of his colleagues, particularly the conservative ones in England, to mistrust his reports of great discoveries. It was therefore one of my aims to try to determine whether in fact his reports tallied well with the field evidence and, in some detail, whether his collecting was tied in with

precise and trustworthy field data. I here say at once, as I did at the time
to those who raised the question, that I found that his field operations
with his wife Mary met the highest professional standards at that time
and that while one might well differ on some points of personal judgment, expressed as such, his statements of fact were indeed factual. It
still seems to me unfair, indeed little short of scandalous, that a few
carping critics prevented his election to fellowship in the Royal Society.
(Foreign members, like me, cannot vote in those elections.)

Not long before this, Louis had received a few fossils from a farmer,
Fred Wicker, near Fort Ternan in what were then known as the White
Highlands between the Rift Valley and Lake Victoria. The few scraps
suggested a fauna possibly distinct from others then known in East
Africa. Mary did not think much of the prospect, but I had encouraged
Louis to investigate at least a bit further and he had sent his black foreman, Heselan Mukili, with a gang of five other African workers. Our
first goal was to check on work there. We found that Heselan had
uncovered a rich bone bed, which now looked so promising that five
more African workers were sent for. We then went back to Nakuru, in
the Rift Valley, the nearest fair-sized town. Louis had to go on to
Nairobi, but we spent a couple of days in a comfortable hotel there.
The great local sights in that part of the Rift Valley, which we saw with
astonishment and delight, are shallow lakes, especially Lake Nakuru
and Lake Elmenteita, where there are literally hundreds of thousands of
flamingos, a phenomenon known to bird watchers throughout the
world.

Another thing I especially recall about our short stay in Nakuru is
that I there carried on a conversation extending over half an hour
entirely in Swahili and I understood every word. I said "Gin and
tonic," which is Swahili for "Gin and tonic." When it came I gave the
waiter a shilling and he said *"Asante, bwana,"* which is Swahili for
"Thank you, sir." Half an hour later I said "Gin and tonic" again and
when it came I gave him another shilling. He said *"Asante, asante,
bwana!"*

A son of Louis's by a previous marriage, John, a herpetologist, came
out to Nakuru and drove us back to Nairobi, We went over the high
(to 13,000 feet) and beautiful Aberdare range, a national park. The park
was closed because of bad travel conditions said to have been caused by
elephants tearing up the dirt roads, which sounded interesting. We had
special permission to enter and so had a whole national park to ourselves. It left nothing to be desired in the way of scenery, but the animal watching was poor. We saw no elephants.

After a few more days in and about Nairobi we drove down to Olduvai Gorge with Louis, R. E. Ewer, and a camp hand. Ewer, known as Griff, is a very able vertebrate zoologist and paleontologist. She was then living in South Africa but shortly thereafter moved to Ghana. On the way we slept on low folding cots in an abandoned coffee plantation on the outer slopes of Ngorongoro Crater. The trees were full of bush babies and they began talking to each other in weird babyish voices as soon as it was dark. It was idyllic until about 4 A.M., when I awoke to find myself covered with hundreds of fiercely biting ants. I tore off all my clothes and managed to get rid of most, but not until morning all, of the ants.

Next day we reached Olduvai, which was deserted and seemed about as remote a spot as one could imagine. I have heard that it is regularly visited by tours now, which seems a pity in a way. No one was working there, but the Leakeys had left a sunshade or open shelter at their most recent (seventh) camp, and we moved into that. The Leakeys have collected a whole series of successive faunas from the sides of the gorge, including remains of at least four distinct species of prehistoric hominids (members of the same family as ourselves). The geology is quite complex, and it was particularly important that I check it thoroughly. With much help from Louis, I did so. With excusable but now not completely accurate enthusiasm I wrote:

"Louis has now also found here an absolutely unparalleled sequence of remains of the creatures who made or used the artifacts. The earliest ones, from the very bottom, are more ape than human. The latest are *Homo sapiens,* even as you or I. There cannot possibly be a more historic spot anywhere on earth. *That* is why we were there, to marvel and admire, and to feel humble before the long struggle of our ancestors."

We had one more trivial adventure. Louis, Griff, Anne, and I were down right at the bottom of the gorge when Louis said, "Hello! There's a tick bird. There must be a rhino near." We took a couple of steps and looked again into the thorn scrub. The rhino was not near. He was *right there!* We retired modestly, each considering what he or she would do if or when the rhino charged. We got up the slope, and looking back saw that the rhino had also retired modestly, up the opposite slope.

On our way back to Nairobi we stopped over at Lake Manyara and at Amboseli. The most memorable things at Lake Manyara were a large herd of wild buffaloes (mbogo in Swahili), who were so testy that we gladly yielded the right of way, and a playful gaggle of dwarf mongooses (I don't know the Swahili name), who were sporting in the middle of the road and were so delightful that we gave them too the right of way. At Amboseli most memorable of many memorable things were the elephants (*tembo*); a pride of lions (*simba*), the lionesses coming back from a hunt and checking in with the old man; and an unusual cloudless view of noble Kilimanjaro and its ice cap. It was also rather fun this time, as we were in a car, to meet rhinos (*kifaru*) that did charge us.

After a couple of days in Nairobi we went off to Lake Victoria with Louis, via Nakuru and Fort Ternan again. The work at Fort Ternan had been carried much further. I had gone along the hillside to take a picture of the now large excavation when Louis came running full out toward me yelling, "George! George! I've got it! It's here! I've got it! George! I've got it!" What he had was a fragment of a skull with two molars and some other tooth roots or sockets. It had a sort of protohuman look to it and might help to fill in a long gap in Africa between the early Miocene primates and the hominids known from the early Pleistocene. This was a really exciting moment in the history of paleontology and of anthropology.

Leakey was somewhat prone to give new names to his discoveries on the basis of only slight variation from species or genera previously named. That is one of the matters of judgment, as previously noted, on which one can disagree with him without impugning either his honesty or his accuracy. In this case he gave the specimen a new generic name, *Kenyapithecus,* and also specific name, *Kenyapithecus wickeri* (for the then owner of the farm where it was found). Most later students think that it belongs to the genus *Ramapithecus,* earlier named from specimens found in India, and some of them place it also in a species first named from the Indian specimens, *Ramapithecus punjabicus.* However that may be, there is a consensus that it probably belongs to the human family (Hominidae) and at more than 10 million years is now the oldest known member of that family.

As soon as we reached a telephone Louis called Mary to tell her about the find, perhaps with a bit of "I told you so" because of her skepticism about the value of work at that site. In fact she was still skeptical until Louis assured her that I had seen the new primate specimen and agreed that it was highly important.

In anticipation of celebration Louis bought champagne in Kisumu, the second city of Kenya but then, at least (I have not been back), not impressive. It is a port on the Kavirondo Gulf of Lake Victoria, and there we boarded Louis's cabin cruiser, the *Miocene Lady.* It had a twin-screw inboard motor and accommodations for up to four people plus servants. Rich deposits of Miocene mammals, mostly between 20 and 25 million years old, occur in islands in Lake Victoria, and the Miocene Lady, broad in the beam, getting old, and not too fast (as becomes a lady) was the work boat for those localities.

That evening we anchored just off the shore of the best-known of the fossiliferous islands, Rusinga, and had a gala dinner accompanied by champagne and preceded by some Leakey Safari Specials. The latter are composed of equal parts of condensed milk and cognac, which sounds horrible but is a very effective pick-me-up after a hard day in the field. The champagne, in celebration and not more or less routine like the specials, was warm, and more of it went to the *Miocene Lady* than into us. In the night we could hear drums from all over the island, no doubt signaling our arrival.

There are many fossil localities on Rusinga and others on other islands and along the mainland shores. We spent some time on the lake visiting almost all of these. The whole cruise was delightful, a pleasure enhanced by feasts of tilapia, the curious African fish that broods its eggs in its mouth. We had heard of that strange habit, but we had not previously known that the fish is delicious.

Once we were almost run down by an island. We had anchored, but suddenly the men pulled up the anchor, started the engine, and got under way as fast as possible — an island was rapidly bearing down on us and threatening to wreck us. There are many floating islands in Lake Victoria, mats of vegetation that break off and float away from shore. This one had a couple of erect trees on it that acted like sails in the moderate breeze.

Back on shore, we visited another famous Miocene fossil locality, Songhor, dropped in again at Fort Ternan, and so back to Nairobi where I finished up notes, lectured at Royal College, and gave a farewell dinner for the Leakey family and a couple of other friends. On the way home we stopped for a few days each in Greece and Italy, relaxing and sightseeing, visiting, among many places, Mycenae in Greece and Capri in Italy.

Later visits to Africa were not primarily professional but always did contribute to some of my various vocations and automatically contributed to my avocations because travel itself stands high among them.

In February – March of 1967 we traveled right around the whole of Africa on a regular cruise ship, the Bergensfjord. We were moving from Cambridge to Tucson and this trip was partly to fill in intervening time. We had both been quite ill, and we also hoped that we would convalesce on the ship, a hope not fulfilled. We were astonished to find that a majority of the passengers, whom we called "the cruise bums," lived on the ship, went on all its successive cruises, and had no other homes. One of them explained to us that this was the cheapest, easiest way to have comfortable lodging, trained servants, and good food cooked by somebody else. Many of them were widows whose ideas of passing time and of cheapness were unlike ours.

We stopped briefly in Santa Cruz on Tenerife (Canary Islands); spent Darwin's birthday (celebrated as Lincoln's on the ship) in Dakar; went on an ill-managed shore excursion in Monrovia (Liberia), founded by American exslaves who went back to Africa where they could have slaves of their own; and went past zero-zero, the spot in the West African bight at Longitude 0°, latitude 0°; to Pointe Noire, in what was then called the Republic of the Congo; on to Luanda, in what was then Portuguese Angola; and visited Cape Town and Durban in South Africa. Shore excursions, run by Cook, continued ill-managed — on other occasions, too, we have had uniformly bad service from that travel company and I welcome this occasion to give them a well-deserved nonrecommendation. By air to the Pretoria-Johannesburg airport and a visit to those two cities and to an ugly monument on the spot where, they say, two hundred Boers with guns killed thirty thousand Zulus with spears. I doubt the latter figure, but there is no doubt that it was a glorious massacre. There is a museum there but we were not allowed in because on that day admittance was for nonwhites only. I don't know why a nonwhite would be interested, but indeed if I were a Boer I would not want to mingle with Zulus where my forebears had slaughtered thirty thousand (or fewer, even) of them.

From Johannesburg we went by train to Kruger Park — excellent animal-watching — among many other species thousands of splendid impalas and a large pack of Cape hunting dogs. Those wild dogs are despised and considered "cruel" by people who joyously celebrate the cruel killing of black humans. The dogs' "cruelty" is simply making a living by hunting down antelopes, which are also hunted nearly to extinction by humans outside the reserves. They are strangely mottled dark brown and a color almost orange, the only wild animals spotted in just this blotchy way.

Our guides in black Africa if not uniformly quite white were nonblack and all were past masters of a game I would call How to Prevaricate without Actually Lying. A fair example was a young lady of Portuguese descent in then still Portuguese Angola who assured us that all qualified Negroes could vote and that they could live anywhere if they paid the rent. What she did not say was that the qualifications for voting were different for whites and blacks and practically impossible for any of the blacks to meet, and that no blacks could afford the rents charged for decent housing, or any housing in a nonblack neighborhood. Yet the strongest racism we encountered was in Liberia, reverse racism, to be sure, but racism just the same. There no white could become a citizen or even live in the country unless specifically employed for something no black could, or would, do.

From Kruger we were driven to Lourenço Marques in Moçambique and went on by the ship to Mombasa, an ancient Arab-Negro melting pot. Swahili is the native language of Zanzibar, Mombasa, and adjacent parts of the East African coast. It has incorporated, with phonetic and grammatical changes, a number of Arabic words, especially for things such as books, unknown until brought by Arab traders. The same kind of borrowing happened on a much larger scale with the fusion of thousands of French words into English after 1066. A variety of Swahili is also widely spoken as a sort of lingua franca among the far more numerous inland blacks for whom it is not a native language. Their "upcountry" or "kitchen" Swahili has a smaller vocabulary and is ungrammatical from the point of view of coastal native Swahili speakers.

At Massawa, in northeastern Ethiopia, we were told that no cruise ship had ever stopped there before, and we later had reason to believe that this ship, at least, had not. When the ship attempted to leave it crashed into the pier, breaking off a stanchion, then went aground, and finally limped out hours later. The small city itself was clearly not used to tourists and was wildly picturesque, quite different from any other sub-Saharan place we had visited but very like a North African Berber or Arab town. It is in what was formerly Eritrea, of which it was once the capital. It is predominantly Muslim and differs religiously, linguistically, and racially from central Ethiopia, or Ethiopia proper, with which it is unsympathetic. Some of the women were veiled, but most of those in the streets were not. One small, bold, and probably hopeful wench followed us for a while and asked me in scanty English whether we were married. Many, perhaps all, of the bare-headed women were evidently prostitutes, two of whom solicited me openly, but with giggles, even though I was with Anne.

The cruise bums hated this stop: "A terrible place." "Our worst stop." "Should never have been visited." "Must be complained about."

Specific complaints all boiled down to this: nothing, or at least not enough, to buy and no modern facilities. Why on earth did they ever leave New York, full of things to buy and every conceivable modernity?

Meanwhile kites circled above the city. Pigeons wheeled around in formation. Gulls went their individual and raucous ways. Arcades made walking bearable in the otherwise merciless sun glaring on white-washed, blue-washed, yellow-washed, and unwashed stone buildings. Life went on much as it had in the time of Muhammad.

From Cairo we went by night train to Luxor. With one dining car for an enormously long train we missed out on two sittings and got into the third only by main force and rudeness on the part of our courier, an Americanized Christian Egyptian. There was a screaming battle in Arabic, German, French, and English. One small segment of the unpleasant melee went:

> *Woman:* But we are diplomats! You cannot displace us!
> *Courier:* So what? We are all diplomats.
> *Woman:* Are you Egyptian?
> *Courier* (laughing madly at his own wit): Worse than that! I am American!

We were finally fed.

Luxor, Valley of the Kings, Karnak. I have already written of my fascination with ancient Egypt, and these places, unseen before by me, I found overpoweringly marvelous. Again I do not undertake to describe what has so often been described and pictured.

Back in Cairo, we found that city being leveled out, internationalized, and depersonalized since our last visit. No camels and donkeys in the streets now. Almost everyone in European clothes. Urban renewal replacing the picturesque old with the doubtless more sanitary modern. The old Shepheard's replaced by a new and modern one on a different site and by a Nile Hilton, hideous outside and just like Chicago inside.

At a Coptic (Christian) church supposed to have been erected on the site where the Holy Family had sojourned in Egypt a cruise bum asked how long before Christ was born it had been built.

We rejoined the ship in Port Said and made later stops in Lebanon (then entirely peaceful), a couple of Greek islands and Greece itself, Italy, France, Spain, and Portugal. As far as health permitted, we had a great time in all, but one book is not long enough for details of a long life and this chapter is about Africa. I append only these random and minor observations:

The red sea is blue.

In Luxor a street beggar polished my shoes and a policeman threatened to arrest me when I tried to pay him.

There were two Soviet war ships at Port Said to protect the Suez Canal from attack. From *Russian* attack?

In a Lebanon hotel presumably male and female restrooms were designated by silhouettes so similar that I could not distinguish them, but it did not matter because one could see into both even when occupied.

In Sicily a distinctly Jewish cruise bum was so fascinated by a jew's harp that he bought several.

In 1970 I had business in London and in Stockholm, as I will mention again in due course. Anne suggested that since we had gone that far we might as well come home by way of the Indian Ocean. That comes in here because having already made a bit of a detour we decided to stop in Kenya and Uganda on our way back from the Seychelles. We found Nairobi still pleasant after *uhuru* (Swahili for "liberation"), and Uganda was then a sensible place, not yet ruled by a dangerous madman.

From Nairobi we flew by small charter plane to a comfortable lodge on the right bank of the Victoria Nile a few miles below Murchison Falls, another good entry in our collection of waterfalls. Here we enjoyed the animal watching, including a lady elephant and her elephant child who had the habit of strolling by the lodge every evening (there was a sign reminding visitors that these were wild animals and not safe pets), also thousands of hippos and hundreds of crocodiles in the river and along its banks. Here we had a somewhat more than titillating adventure. We went with a small party of tourists by boat to the foot of the falls, and just there the motor conked out. We were whirled around, dashed against rocks with an awful thud, and then carried erratically downstream to where we circled for a long time in what was appropriately mapped as Hippo Pool. A group — bevy? pride? no, *ponderosity* — of hippos popped up, looked us over, submerged, emerged a bit closer, over and over, agog with curiosity at our strange behavior and, I felt sure, at our ugliness by proper, that is, hippo standards. After an hour and a half the African crew did get the motor started.

Our next and up to now last visit to Africa was one of the best, but it started almost by accident. In fact one could truthfully say that it was taken literally because of an accident. In 1972 we had planned to make a third trip to Antarctica on our favorite small ship, the *Explorer*. On the day we were to fly down to Argentina to board the ship we received word that the *Explorer* was aground in the South Shetland Islands and our cruise was canceled. We were at a loose end, as we had made all arrangements for a long absence and had house-sitting friends in occupancy in our Tucson house. So we decided to go to Africa instead. My sister Martha Eastlake, known as Marty, was already in Buenos Aires expecting to meet us and join the Antarctic cruise, but our ingenious travel agent and friend Wayne Randall got her to Johannesburg by the hitherto unheard of route Buenos Aires-Sierra Leone-Ghana-Johannesburg and us by the orthodox route via London. We arrived less than a day apart.

We had a picturesque, delicious, and hilarious dinner with Philip Tobias, a physical anthropologist and anatomist whom Anne and I had met in Nairobi in 1961. I spent a day in his department at the university, where the first specimen of *Australopithecus* and some other specimens of that group are preserved. Those creatures are now generally accepted as closely allied if not actually ancestral to ourselves, but the first discovery started a generation of controversy and a complex of conflicting interpretations. Philip calls it "the face that launched a thousand slips."

From there we went off on a long tour with a driver-courier. He was a Boer or Afrikaner, country-raised and excellent at spotting and naming birds, less so but still good with mammals. He was nevertheless a difficult character, moody and suspicious to the point of paranoia, and by the end of our association we were glad to see the last of him. He hated English-speaking South Africans, whom he called *pommies,* and had an Afrikaner inferiority complex. He insisted that his ancestry was not Dutch (it was), that his language, Afrikaans, has no relationship with Dutch (it is in fact a somewhat archaic Dutch dialect), and that we looked down on him because he was and we were not Afrikaner (we did not). In general we got the impression that the barrier between Afrikaners and the South Africans of English descent, although less conspicuous, was almost as great as that between whites and blacks.

From South Africa we crossed into Rhodesia at Beitbridge, across what Kipling called "the great, grey-green, greasy Limpopo River." The water is not grey-green, more just dirty, and Kipling probably never saw it, but his alliteration lingers with me from childhood. In the Rhodes Matopos National Park we climbed the great granite *koppie* (Afrikaans for *kopje,* "hill") on top of which Rhodes is buried. To Anne's surprise I took off my hat to him: he was a robber-baron sort in a way but also a great man in other ways. Driving back, we saw a group of dassies (hyraxes or conies, ridiculously "jack rabbits" in our driver's imperfect English) on a height above the road. We collectively parodied:

> A dassie on the koppie's brim
> *Procavia capensis* was to him
> And it was nothing more.

One of many new species for us on this trip.

We detoured into Wankie National Park and one of the new species for us there was a blue wildebeest. Wildebeests are gnus and I think there should be a song in praise of "blue gnus," but inspiration did not come.

A stay at Victoria Falls sensationally completed our acquaintance with the Great Waterfalls of the World.

We then went to Kariba at the dam that forms what is claimed to be, and may in fact be, the largest artificial lake in the world. Victoria Falls, and this dam, far downstream from the falls, are both on the Zambesi River, which here forms the boundary between what had been Northern Rhodesia, already Zambia at the time of our visit, and former Southern Rhodesia, now simply Rhodesia, and beyond any doubt soon to be Zimbabwe. I wonder whether Rhodes's body will then still rest in the Matopo hills. (His name will continue in the Rhodes Scholarships, at least.) The great dam at Kariba was built when Northern Rhodesia and Southern Rhodesia were still united and it supplies hydroelectric power to both Zambia and present Rhodesia. When we were there the Zambians were installing their own generators on their side of the dam in order to end any further dependence on Rhodesia.

We visited and did not like Salisbury, capital of Rhodesia, just another city with no distinction. We then visited and did very much like Zimbabwe, the ancient ruins whose name will become that of the whole country when it comes under majority rule. This is just the greatest of a number of ruins in this general area comprising high walls made of granite spalls in erratic patterns. The preserved walls were not parts of dwellings and it is presumed that they served for defense in a peculiar, perhaps also ritualistic way, and that the dwellings, long vanished, were of wood or wattle daub with thatch. In any case the ruins are tremendously impressive and bespeak a lively, mysterious, long vanished culture. While there we also visited Lake Kyle, on the Mtilikwe River, and for the first time saw "white" rhinoceroses.

The country lodge where we stayed near Zimbabwe was downright luxurious, with white owners, of course, and a large staff of well-trained black servitors. The proprietress gave us a long farewell talk. She wanted us to let everyone back home know that Rhodesia was a completely "happy country"; in fact even then outlying white settlers were living in a state of siege and blacks went in fear of their lives. She also assured us that the proprietor and she could never go off and leave "our children" — she meant their black servants — "defenseless." It had never entered her head that the "children" were grown men with aspirations and capabilities of their own. Her attitude was exactly what might have been expected of an antebellum southern planter in the United States who was kind to his dogs and his slaves as long as they obeyed him.

We thought of that some days later when we drove through the "independent country" of Swaziland. Our white Afrikaner driver said scornfully, "We just let them think they are independent." There are large asbestos mines there owned by white South Africans and we passed a company town with three quite separate areas, one of luxurious houses and gardens for company officers (all white), one more modest but comfortable for office workers (also white), and one of crowded, squalid huts (you know whom they were for). We verified an interesting mathematical relationship: the number of Swazis sharing in the management and profits of commercial mining, timber, and other such enterprises in "their" country is exactly equal to the number of whites sharing in the physical labor involved. The number is zero, of course. So why did South Africa grant the Swazis "independence"? Well, one interesting point among several is that South African whites cannot own property in black reserves, but of course they can do so in an "independent" country.

On our way south we stopped at Kruger Park and also Hluhluwe. Kruger was again delightful. The most unusual thing there was that very early one morning we saw two young spotted hyenas frolicking about in a water hole, playing like two kids, splashing and ducking each other until they had had enough and ran off together. It is not customary to think of hyenas as endearing, but these were. I don't really understand the dislike for hyenas. Even a standard dictionary defines them as "cowardly." They are not as cowardly as the men who kill them. In Hluhluwe, where to our annoyance tourists must put up at a Holiday Inn, the special attraction is "white" rhinoceroses, which we saw from far and near. Farther along toward Cape Town we encountered a string of lorries, each with a big crate containing a "white" rhino. These great creatures, exceeded in size only by elephants, have disappeared from most of their former range but are doing so well in some of the reserves, especially Hluhluwe, that it is both possible and desirable to thin them out by transfer to other game parks and zoos.

Eventually we reached Cape Town, where we rested for a few days, and then drove back to Johannesburg by way of Kimberley, among other places. That took us right across the Karoo, an arid stretch of Cape Province. Its name, also spelled Karroo, does not appear on ordinary maps but is well known to all paleontologists and to most other geologists in the world. In a tremendous series of sediments there are abundant fossils, mostly reptiles, of the Permian and Triassic periods, roughly 200 to 250 million years ago. Of special interest among them are hordes of mammallike reptiles from some of which the true mammals, eventually including ourselves, arose.

In Kimberley we took a standard "tour of the diamond mines," which did not include any view of a mine but glimpses of a mill and of the prisonlike compound where the black miners are locked in, treated, and worked like convicts. A South African woman there said, "We do all this for Them" — blacks are generally called Them — "and They don't really appreciate it." On our own we also saw the Big Hole, the first great diamond mine here, now abandoned and half full of water. That was impressive, and the open-air museum alongside is interesting. Several million *pounds* of diamonds, not carats (there are about 227 carats in a pound), were taken out of that one hole. Diamonds are expensive not because they are rare. In fact they are more abundant than many other kinds of gems on the market.

In Johannesburg we found that planned travel to Botswana was impractical because of heavy rains there, and so we flew down to Cape Town. There I enjoyed meeting Brett Hendey, the vertebrate paleontologist at the South African Museum, with whom I had corresponded about fossil penguins. He took me on an excursion to Lambert's Bay, on the Atlantic coast well north of Cape Town, where I could see live penguins and other sea birds.

We finally went to England on the Union Castle Line's *Pendennis Castle,* a fine ship well managed, and flew home from Heathrow.

Our overall impressions of Rhodesia and South Africa in 1972, which we see no reason to alter now, were:

1. Few countries compare with these in scenic, floral, faunal, and geological interest.

2. There are some nice, decent people of any race or color.

3. The legal systems and social mores imposed on these countries by white minorities are so cruel, hypocritical, and disgusting that it is an absurd mockery to call them civilized.

Notes

§   In 1961 most of the shops in Kenya were owned by East Indians, who were relatively prosperous. That is no longer true since *uhuru* (independence).

§   Shirley Coryndon later worked at the British Museum (Natural History) in London, married another British vertebrate paleontologist, R. J. G. Savage, of the University of Bristol, and to the lasting sorrow of her colleagues and friends died in 1976.

§   I saw Louis Leakey often after 1961 in England and in the United States but never again in Africa. I last saw him in 1971 at the Alioto symposium on Teilhard, previously mentioned. He was then obviously failing, and he died in 1972. His wife Mary and one of his sons, Richard, are carrying on his work in Kenya and Tanzania.

§   Although I must go along with authorities in just this field, such as Simons, cited below, I have some lingering doubts about the exact placing of *Kenyapithecus* = *Ramapithecus* in our family tree. The original specimen is now almost always shown, as in figure 109 of Simons's book, with a canine tooth attached, and the size and shape of that tooth are part of the evidence for the human affinities of the specimen. But I can testify that the tooth was not in or near the specimen when it was found, and when I later tried to fit the canine into the broken socket on the original specimen I thought the fit was questionable.

§   In 1972 we did get into the museum on the site of the massacre of the Zulus. We were surprised to find it quite interesting, not as a monument, but for the insight it gave into pioneer life in South Africa.

§   Moçambique is now spelled Mozambique and in 1976 the name of Lourenço Marques was changed to Maputo. One hardly knows from day to day what the current names of some African countries and cities are.

§ "White" rhinoceroses, *Ceratotherium simum,* are quite different from "black" rhinoceroses, *Diceros bicornis,* but both are of nearly the same color, neither white nor black but a sort of dirty grey. There are plenty of guesses as to how they came to be called "white" and "black" but no evidence known to me. "White" rhinoceroses have wide snouts, graze, and are placid when not breeding. "Black" rhinoceroses have pointy snouts, browse, and have lifelong temper tantrums.

References

Cole, S. 1975. *Leakey's luck: the life of Louis Seymour Bazett Leakey,* 1903–1972. New York: Harcourt Brace Jovanovich. An excellent, detailed biography.

Leakey, L. S. B. 1974. *By the evidence: memoirs,* 1932–1951. New York: Harcourt Brace Jovanovich. This posthumous second and last volume of Leakey's rather deadpan autobiography covers some of his most important years in paleontology.

Simons, E. L. 1972. *Primate evolution.* New York: Macmillan. Two of the hominid-like specimens from Fort Ternan are shown in figure 109.

Simpson, G. G. 1975. Notes on variation in penguins and on fossil penguins from the Pliocene of Langebaanweg, Cape Province, South Africa. *Annals of the South African Museum* 69: pp. 59–72. The most recent of three papers on South African penguins, all thanks to Brett Hendey's cooperation.

20

More Changes

Diversité, c'est ma devise. (Diversity, it is my slogan.) — Jean de La
Fontaine

*They are the weakest minded and the hardest hearted men, that most love
variety and change.* — John Ruskin

As with opinions on music, those on variety in one's life are radically
divided. I can only hope that my approval of La Fontaine's slogan does
not qualify me for Ruskin's contempt. However that may be, I now
come anew to circumstances that forced variety upon me, and again I
enjoyed the results although I might not have chosen all of them.

From 1959 to 1967 we lived in apartments in Cambridge, still spending some time each year in our New Mexico home although I was no
longer doing field work in the San Juan Basin. That work was sus-
pended in part because of my disability but mainly because it was considered that the American Museum, which had all my specimens and
field notes, had priority to continue what had been my program there.
That was not in fact done and recently the University of Arizona has
been working there but under my colleague Everett Lindsay and not
directly under me.

Both Anne and I have been ill with one thing or another or a combination of several for most of our lives. The details are of no real interest
even to us, and I mention illness only when it had a direct and important effect on our lives. That happened in 1964 when we were in New
Mexico. We came down with his and her heart failures and were hospi-
talized in Albuquerque. There was a row because we were separated
and could not check on each other. The hospital management consid-

ered it scandalous to have people of opposite sex in the same room even if they had been well and truly married for years, were in separate beds, and too ill to do anything interesting anyway. Our doctor finally prevailed, with the argument that we worried so much about each other that we would never get well if kept separate. When we were able to get back to our mountain home, Los Pinavetes, Anne's sister Pat and her husband Les Oldt spent their vacation giving us tender loving care. Toward the end of the summer I was back in the hospital more briefly and Anne in a nursing home, where I soon joined her. As soon as we could get about at all the doctor recommended a longer convalescence and on his advice we decided to take it on a sea voyage across the Pacific and back.

That was the first lengthy travel I had ever taken without any directly professional aspect. We have pleasant memories of it, but no journal because for the early part of the voyage I was still too ill to start one. I will treat the trip only briefly here. We went out from Los Angeles on the P. & O. liner *Oriana* with stops in San Francisco, Vancouver, Honolulu, Suva (Fiji), Auckland, and Sydney. We lay over for about a week in Sydney and then returned on the P. & O. liner *Oronsay* via Manila, Hong King, Yokohama, Honolulu, and Vancouver to San Francisco, from which we flew back to Cambridge, able to resume work there. Those ships stand out in memory along with the Castle liner mentioned in the last chapter as the only large ships that we found thoroughly agreeable and well-run. Alas, none of them is still in service.

We had brief visits with relatives in Honolulu, liked Fiji very much and decided to come back, which we did; also liked what little we saw of New Zealand and decided to see more of it later, also done; and visited with friends from our earlier visit in Sydney. Especially pleasant was a picnic with Charles Birch, professor of genetics at the University of New South Wales. In a national park not far from Sydney Charles supplied and grilled chops and we supplied pink champagne, a farewell gift from my sister Peg in Los Angeles, saved because I could not drink it earlier. We saw and heard our first lyre birds in a forest nearby. Their vocalizations are even more varied and extensive than those of our mockingbirds.

The *Oronsay,* smaller and older than the *Oriana,* was even more to our taste, being more homely in the British sense of the word. In Manila I admired the pretty Philipinas and we took the practically obligatory tour out to the Taal volcano. In Hong Kong we followed the ritual of having some quick tailoring done and had a splendid meal on a floating restaurant. From Yokohama we had a long evening in Tokyo and an overnight trip into the mountains, including a boat trip the length of Lake Hakone. We were in fog and rain almost all the time and did not even see Fuji until we were sailing away, when at last we had a fine but distant view. Back in Vancouver we admired the flowers and the hills north of the city and across the harbor.

So we settled down for three more years in Cambridge, Massachusetts, broken only by short separate visits in 1965 to Cambridge, England, and to Paris, France, which is quite easily distinguished from Paris, Arkansas; Paris, Idaho; Paris, Illinois; Paris, Kentucky; Paris, Maine; Paris, Michigan; Paris, Mississippi; Paris, Missouri; Paris, New York; Paris, Ohio; Paris, Ontario; Paris, Tennessee; Paris, Virginia; and especially Paris, Texas. I may also mention that there is a Paris Crossing in Indiana, that the two tiny settlements on Christmas Island in the Pacific are called Paris and London, and that there is a Mount Paris on subantarctic Campbell Island, which has a research station but cannot boast of a settlement. Anne and I went together to Cambridge, where we had a pleasant stay at the university first and then with my colleague F.R. ("Rex") Parrington. Anne was unable to go to Paris with me, which was a pity as I had a wonderful week there as the guest of the University of Paris (generally called the Sorbonne, after one of its oldest buildings) and with much hospitality from colleagues and friends. I did have to sing for my supper, so to speak, to the extent of giving a lecture at the university. I was criticized by the students for speaking in English rather than French, but I was explicitly asked by the university authorities to do so. (Probably they did not like my accent in French, which passes for Parisian in southern France but not in Paris.) That was the last time I saw Paris, which made me regret all the more that Anne was not with me. We have been back to London and some other European cities several times since then but not to Paris.

I have already mentioned some of the more important things I was doing during the last years at Cambridge, Mass., including publication of several books. I also thoroughly revised *The Meaning of Evolution* and wrote many papers, including, for example, two on racial problems from a taxonomic point of view done for UNESCO and published in English in the *International Social Science Journal* and in French in the French version of that journal. I also continued to lecture hither and yon, for example giving the Thomas Burke Memorial Lecture at Washington State Museum, published there as "Mammals around the Pacific." Although it means being nonchronological again, I should also mention that Anne and I attended closed conferences in 1961 and in 1974 at the Wenner-Gren Foundation's wonderful castle-aerie Burg Wartenstein in Austria. At the earlier conference on classification and human evolution I held forth on the meaning of taxonomic statements and Anne did so on psychological definitions of man. Paul Fejos, a great man, was cheerfully there although fatally ill. At the later conference his widow, now Mrs. Lita Osmundsen, managed gracefully and efficiently. That conference was on phylogeny of the primates. I spoke on recent methods of phylogenetic inference, in addition to making, not for publication, some remarks on phylogeny and classification that were well received there but caused me to be almost insanely denunciated when an unofficial taping of them was played at my former institution in New York. Anne also spoke but did not prepare a paper for publication. If a taping of her remarks was also made and then played elsewhere, we were spared the reaction.

In 1967 we made our next big move. We had anticipated eventual retirement by buying a house in Tucson, but we had not intended to move until 1972 or so and our house there was rented in the meantime. However, Anne was having pneumonia every year in the vile Cambridge winters; a heart operation had not appreciably helped her cardiac problems; and I was not exactly in the pink either. It seemed that if we were to enjoy our western home we had better start doing so as soon as possible. Anne retired as professor emerita from Harvard. My appointment as Alexander Agassiz Professor in the M.C.Z. was renegotiated to run at least until 30 June 1970 but on a half-time basis and with half pay, and with the condition that I continue to do research as from the M.C.Z. and spend some time there each year. We moved our household to Tucson, where Anne was appointed (unpaid) lecturer in the psychology department of the School of Liberal Arts and I was appointed half-time professor in what was then the geology department of the School of Mines but became the Department of Geosciences of the School of Earth Sciences. Until my appointment on the M.C.Z. staff ended in 1970 and was not renewed, I did regularly return to the M.C.Z. and Harvard, usually staying at the Faculty Club. I continued active research, which until July 1970, was divided between the M.C.Z. and the University of Arizona. Visits at Harvard were not always pleasant because that was a period when revolutionary activists were attempting to close the university permanently. The young hoodlums did succeed in making nuisances of themselves, among many other ways by invading the Faculty Club and by causing its evacuation with bomb threats. While agreeing that most institutions, including Harvard, can do with a reassessment and a shaking up from time to time, I did not and still do not see what possible benefit to anyone could have come from those merely destructive activities.

In Tucson a much younger vertebrate paleontologist, Everett Lindsay, joined the university department at the same time as I. He is a most congenial colleague who has taken the basic responsibility for formal teaching and our two families have become good friends. We also met again some old acquaintances who now became close friends, notably Lawrence ("Larry") Gould and his wife Peg. Larry, a noted Antarctic explorer and former president of Carleton College, is now also a professor in the Department of Geosciences here. We have also been fortunate in making new friends here and although we do still miss some of the old friends in New York and in Cambridge we do not lack for pleasant social as well as professional contacts.

Since 1970 I have had no connection at all with the M.C.Z., although I am still in touch with some of the individuals there. Harvard made me a professor emeritus and keeps me informed of university activities. I have continued at nominally half time in the University of Arizona, which is convenient since with some exceptions it permits me to make my own schedule. As I was overage almost from the start, I have been reappointed a year at a time by special action of the State Board of Regents. The department is large, well-staffed, and productive, but it and the university as a whole lack extensive facilities in my own fields of study, and that has led to some reorientation both physical and mental. On the physical side it was soon obvious that I could not set up adequate research quarters in my cramped, overcrowded department on the campus, so we built a separate building, with one large room, in the patio of our home. There we installed our extensive professional libraries and have desks, work tables, and other facilities for Anne, me, and a part-time secretary employed by us, not the university.

In connection, with that reorientation we set up the Simroe Foundation, its name a somewhat simpleminded derivative of our surnames. This is chartered by the state as a nonprofit foundation and after investigation has been accepted by the I.R.S. as tax-exempt. Its purpose is to make our technical libraries and research facilities publicly available to qualified students and eventually to transfer them to other nonprofit and public insititutions where they will be permanently and still more widely useful. The foundation has no endowment, makes no grants, and has no paid employees. It is administered by a board of trustees, at present composed of three paleontologists, two psychologists, a biologist, and an attorney. Although Anne continues some work in her fields, she is less active than formerly and is technically retired, so with her acquiescence and by vote of the board most of her library was presented to Howard University. My library is still in quite active use by faculty and students of the University of Arizona and some others and so remains in the annex on our property. It is also used for reference in connection with frequent requests for information and for refereeing of manuscripts and of grant proposals, many of which are referred to us.

There was also some mental reorientation toward the kinds of research that I could effectively do after the change in headquarters. I continued to do what are called "overviews" in contemporary jargon — for example, one on the first three billion years of community evolution prepared for a Brookhaven symposium — and extensive essays on special topics — for example, one on Mesozoic mammals for a symposium at the Linnean Society in London and one on the concept of progress in organic evolution for an issue of *Social Research* devoted to various aspects and problems related to progress. For the most basic research, that is, the description and classification of fossils, I have relied largely on borrowed specimens and on visits to other institutions.

I had planned a monograph on the fossil marsupials of North America and many specimens were made available to me, but I was refused access to the crucial collection at my old institution, the American Museum of Natural History, and I therefore had to abandon that plan. I was nevertheless able to continue work on South American fossils with special emphasis on the more varied marsupials of that continent. (The study on the argyrolagids, mentioned in chapter 8, was done in this period.) I also continued study of penguins, fossil and recent. I have already told (chapter 14) how I came to work on fossil penguins and that my interest extended to the living species as well. This study has continued and culminated since I moved to Tucson. Until quite recently I was the only student of fossil penguins, and I coined a term for this bizarre and exclusive speciality: among other things, I am a paleospheniscologist. That reminds me that once when one of our daughters was being a bit rambunctious I suggested that she might show some respect for the world's greatest paleospheniscologist. She countered, quite correctly, "Yes, Papa, but also the world's worst."

Studying fossil penguins and watching recent ones has involved me in wide travels, some of which will be my next topic, but first I have to make some perhaps ill-natured remarks on the subject of bird watching and birding. There are thousands, perhaps even millions, of people who engage in what they now prefer to call "birding." The aim is to see a bird, identify it, and enter its name in a life list. Period. In true American (or British) spirit, getting a bigger life list has become a highly competitive sport. The world champion at present is said to be an Englishman, moved to the United States, with a life list of 5,340 species — he still has some to add, for Ernst Mayr, the professional authority in this field, estimates that there are 8,600 species of birds extant. Still 5,340 is enough to get the champ into such company as that of the heaviest woman, the largest nudist camp, the most alcoholic person, and the world's longest apple peel as recorded in the *Guinness Book of World Records*. Sight identification is the purpose of the Peterson bird guides. I use them, too, and I have enjoyed voyaging with Roger Peterson, but I *watch* birds in addition to identifying them and I am glad that the identifiers prefer to be called "birders" and so leave the designation of "bird watchers" to people more like me.

With that I cut this chapter short. I will devote parts of the next two chapters to some forays into penguin country.

Note

§ It is proper at this point to give one example of the reasons for my increasingly ambivalent attitude toward the M.C.Z. During our illnesses and convalescences in 1964, Anne, employed by Harvard, was given sick leave on full salary. I, employed by the M.C.Z., was put on leave without pay. It would be unpleasant to go into further detail and to give other examples. It must suffice to say that on the administrative level I was treated badly in the M.C.Z. On other levels I enjoyed my work and personal relationships in that institution.

21

Antipodes

Hurried my soule to the Antipodian strand. — Richard Brome (1640)

The antipode of Tucson is in fact a point in the Indian Ocean very far from any land and roughly half way between the southwestern end of Australia and the African coast somewhat north of East London in Cape Province. Nevertheless it is part of our English heritage that we usually think of Australia and New Zealand as the antipodean countries. In 1968, a year after our move to Tucson, we made not our first or last but our longest and in some respects our most interesting journey to and in those countries. Our reasons were professional in several different ways, but we also had the personal motives and satisfactions that travel blends so nicely with professional aims.

On our way in September we stopped at the annual meeting of the American Psychological Association, where Anne, looking both erudite and beautiful as is her wont, accepted the Richardson Creativity Award. We then stopped for some days on Viti Levu in Fiji, flying by small plane from the international airport at Nadi to Suva at the opposite (eastern) end of the island. While exploring the town and other nearby places we stayed at the Grand Pacific Hotel, or the G.P. as we old Fiji-hands call it, a charming old-fashioned hostelry. I am glad to say that it survives in 1976, the charm only slightly diminished, although Suva now boasts characterless American chain tourist traps. Even in 1968 we were distressed that the Fijians, who had earlier and becomingly had what are now called "Afro" hairdos in the United States, although we never saw one in Africa, had abandoned them for shorter cuts. And even then the great attraction of the island and of the congenial Fijians was diluted by the unpleasantness of the East Indian storekeepers, taxi drivers, peddlers, and wheeler-dealers.

We drove back to Nadi by the Queen's road, along the southern shore (the northern road is the King's), and flew with a change of planes in Sydney to Melbourne. There we enjoyed for some time the hospitality of Edmund Gill, then deputy or assistant director of the National Museum of Victoria, and of his charming wife Kath. I worked in the museum going over the fossil mammals. I also found there some undescribed and unidentified fossil penguins, which could not then be studied in detail but which later were generously sent on loan and published by me. At nearby La Trobe, then the newest of Australia's rapidly expanding universities, I was able to see live *Notomys* and *Antechinomys,* creatures of special interest to me. *Notomys,* a placental mammal, member of the rat family (Muridae), confined to Australia, is bipedal and is remarkably convergent in habits and gaits to our southwestern "kangaroo rats" and "mice," also placental and not kangaroos but likewise neither rats nor mice. *Antechinomys* is a marsupial and had been supposed likewise to be bipedal but was shown (by David Ride) not to be. These observations formed useful background for later studies of the small, extinct South American marsupial bipeds, mentioned in chapter 8.

I gave some small repayment for hospitality by lecturing for Melbourne University and the Royal Society of Victoria. (Victoria is the state of which Melbourne is the capital.) Then we drove with or rather were driven by Edmund on an extensive excursion, primarily to see fossil localities. We often stopped in motels and found them on the whole rather more pleasant than most American motels. One superiority was that in many of them breakfast is included. It is ordered the night before and is delivered at the desired time through a hatch into the room so that one doesn't even have to throw on a dressing gown to get it and take it back to bed. We did not sample one of the usual breakfast dishes: spaghetti on toast.

We went up to Mildura, a thriving town on the Murray River in northwestern Victoria. The Murray, longest river in Australia and the only one of much consequence on that rather dry continent, rises in the Snowy Mountains, or so-called Australian Alps, the only considerable mountains in that low continent, and angles away from the coast for much of its length before finally turning south to the sea in the state of South Australia. For most of its course it forms the boundary between Victoria and New South Wales. Around Mildura its waters irrigate rich land planted mostly in citrus fruits and grapes. Some wine is made, but to our taste it is inferior to the wine of neighboring South Australia. Most of the grapes of Sunraysia, a promotional name for this general area, are dried and sold as raisins.

Sir Robert Blackwood, chancellor of Monash University, and Ken Simpson (not a relation of mine), on the staff of the National Museum of Victoria, met us in Mildura. We crossed the Murray into New South Wales and went to Lake Victoria. In lower levels of the rim of the lake basin there are Pleistocene beds, and there we saw an articulated skeleton, not yet excavated, of a large extinct marsupial. Higher beds have pre-European aboriginal remains and artifacts. We went on to Renmark, a town in southeastern South Australia, where after spending a night in an odd but comfortable municipally owned hotel Sir Robert and Ken left us to go back and camp. There was a project to build a dam on the Murray a bit above Renmark and they were carrying on archaeological work to salvage material in areas that would be flooded. After some agreeable further visits around the rich fruit-growing areas on the Murray, Edmund and we went to Wyperfeld National Park.

That national park, back in Victoria, is the largest in the state, although its 218 square miles do not make it enormous. It was then cared for and protected by one part-time ranger who also farmed nearby in order to eke out a living. There were quite a few campers in the park, and perhaps that is why even with the earnest guidance of the ranger we saw no ground mammals or birds beyond mobs of kangaroos — the big gray ones (*Macropus*) — and emus. We did see large numbers of colorful flying birds, almost all of the parrot family: galahs (a kind of parrot; the second syllable is accented), cockatoos, budgerigars, ringneck parrots, and what the ranger called "smokers," more formally called regent parrots.

From Hamilton, a good-sized town in southwestern Victoria, we went out to Grange Burn, one of the still sparse pre-Pleistocene mammal localities in Australia.

Between Hamilton and Melbourne to the east there is a hilly, open, lake-strewn region almost all of which has been cleared and planted in foreign grasses. This gives it a pleasantly green and pastoral aspect, but I could not help thinking that it had involved annihilation of the native flora and most of the native fauna. Here we had the delightful experience of staying at Talindert, the homestead of Sir Chester and Lady Manifold. Americans can be quite taken aback by what Australians call a "homestead." The name was doubtless appropriate even in the pioneer American sense when squatters first moved into the out back and took possession, but that is now several generations ago — in the case of Talindert four or five before Sir Richard; he professed not to be quite sure which. Now, as here, homesteads are likely to be the luxurious homes of the aristocrats of an area. Talindert is enormous. It was built in the 1890s, with innumerable very large rooms, and it had been reasonably modernized since. We slept in Princess Alexandra's Room, that is the one in which she slept when at Talindert. We could not gather whether her occupancy was unique or habitual; the room was quite feminine in a regal sort of way. There were magnificent gardens, maintained by three gardeners. Sir Chester protested that he was too old and heavy to do much riding, but he also protested, with a naïveté that was clearly more jocular than real, that he had more thoroughbreds than he really wanted — some forty of them. His excuse was that he sold them with a racing guarantee and when they failed to win a race back they came. It occurred to us that those charming hosts with their interesting name and in their beautiful surroundings should be in a novel written in collaboration by Thackeray and Waugh.

There is a good Camperdown cheese local to that region, which is remarkable because most Australians seem to prefer Kraft processed cheeses.

At the last census there had been four aborigines in the whole state of Victoria, but aboriginal names linger on. Thus the fossil locality that we had come to see near Talindert is named Lake Weeranganuck.

We returned to Melbourne and on 29 September flew to Adelaide. The several days there were full of interest. It was, for instance, then that we revisited Lady Mawson, as mentioned earlier. I shall, however, be more specific only about what is known as the Ediacara fauna, specimens of which I could examine through the courtesy of the paleontologist at the university, Professor Martin Glaessner. The fossils of that fauna were found in rocks approximately 650 million years old in the Ediacara hills, in the central part of eastern South Australia. Their great interest is that they are the oldest surely identifiable true animals, *animals* being defined as multicellular with some differentiation of organs. Much more primitive fossils, akin to bacteria and to certain algae, are known from rocks as old as about 3 billion years, but next to the origin of life the progression to true animals is perhaps the most crucial event in the long, intricate history of organisms, and it is attested by these impressions in the sandstones of Ediacara. Somewhat similar fossils of about the same age have now been found widely scattered and it is probable that faunas like that from Ediacara became world-wide.

There is a curious anecdote about one of those fossils, the one called *Tribrachidium*. This is the enigmatic impression or mold of a creature with three hooked arms radiating from the center of a circular, rayed disk. It is so pretty that a brooch with that pattern was made for Mrs. Glaessner, but she found it inadvisable to wear it. Someone with more imagination than sense or perception mistook it for a variety of swastika and concluded that it was a Nazi symbol.

Here began a strange series of events which made our circuit of the continent almost as memorable to the Australians as to us and suggested that we might be subsidized *not* to visit places. While we were in Adelaide it snowed there for the first time in years, even though in October winter was nominally over. At our next stop, Perth, it snowed again, also for the first time in the memory of many of the inhabitants. When we drove from Perth to Kalgoorlie a town we had just passed through was leveled and the road made impassable by an earthquake. When we reached Alice Springs the railroad station burned down. (Yes, there is a railroad from Adelaide, or more precisely from Port Augusta, to that settlement in the dead heart of Australia.)

A train buff from childhood and now unable to travel on American trains with pleasure, I had long wanted to take the long train trip from Port Pirie in South Australia across the great Nullarbor Plain to Kalgoorlie in Western Australia. Now I could, as the greater part of a journey from Adelaide to Perth. The railroads in Australia were originally built with different gauges for each state and for some branch lines, maliciously, I believe, to force changes of equipment and transshipment of freight. Thus in 1968 we still had to change at Port Pirie on our trip from Adelaide, although still in the same state, and again at Kalgoorlie for Perth, both in the same state. The rebuilding to a wide standard gauge throughout was under way and had gone that far. I believe it has now been completed.

The commonwealth train for the greater part of the trip was already all wide gauge and our compartment was excellent, complete with toilet and shower. Meals, included in the fare, were below the Australian average, which itself is not exactly gourmet, but they sustained life. The Nullarbor Plain is a feature impressive from the very fact that it is so vast and so nearly featureless. Hundreds of miles with no change, absolutely flat, treeless, with low salt bush and blue bush a foot or so high, in good form for us as there had been a recent rain. The rainfall here is always low but the effective aridity is increased by the fact that the plain is a limestone sponge into which such water as does fall sinks at once. No road follows the railway and the "stations" are widely spaced shacks for track workers. Toward the western stretch of the plain on the second day a few scattered eucalypts began to appear and after sunset we pulled into Kalgoorlie. The ride on the extremely cramped narrow gauge train to which we changed there was so rough that it was hard to sit up, let alone stand. It took fourteen hours overnight to run the 380 miles into Perth. But do not be discouraged if you also contemplate this great rail trip. That stretch, too, is now comfortable commonwealth gauge.

At Perth we were met by W. D. L. Ride, who goes by his second given name, David, then director of the Western Australia Museum, now director of the Australian Biological Survey in Canberra, and by Mike Archer, then a student at the University of Western Australia and now on the staff of the Queensland Museum in Brisbane. David is an eminent mammalian systematist and Mike an excellent vertebrate paleontologist. We spent only a couple of days in Perth, divided between work at the university and the museum and some sightseeing. Perth is an attractive city, or a conglomeration of nominally distinct but actually continuous towns, with flowery parks and much water along the shallow, sinuous estuary of the Swan River.

On 7 October we took off on a wonderful drive with Dom Serventy, a naturalist with the CSIRO, which stands for Commonweath Scientific and Industrial Research Organization and is pronounced more or less see-ess-eyeró. No one knows Western Australia better than Dom (christened Dominic), and he recognizes all its birds, near or far, either by sight or by sound. I must confess that except for penguins, southwestern U.S. birds that frequent areas where we have lived, and some of the more colorful families such as kingfishers and parrots, I am usually a somewhat indifferent bird watcher. I did greatly enjoy seeing them with this great authority, but I confess that I was even more entranced by the flowers we encountered in driving southward near or along the shore to the extreme southwestern point of Australia and then looping back farther inland to Perth. We had the good fortune to be in southwestern Western Australia at the height of the flowering season, and the display was an exciting delight. I am not so much an indifferent as an ignorant botanist, but I am a devoted flower watcher. I have seen the flower carpets of our southwestern deserts in the rare years when a spring rain came at just the right time, and in younger days when I could climb high I was enchanted by summer meadows above timberline, but I have never seen a display to equal that around us on our drive with Dom.

To foreign eyes one of the most striking features of the Australian flora is that many plants, mostly large shrubs or trees, have flowers with inconspicuous petals or none at all. The showy parts are formed by many filamentous or spiky brilliantly colored stamens. That is true of the eucalypts, omnipresent in Australia except in the treeless deserts and comprising literally hundreds of species. It is also true of numerous other plants, notably the bottlebrushes of the genus *Callistemon* and several related genera in which the flowers do indeed have the shape of bristly, cylindrical brushes. A *Callistemon* on our place in Arizona blooms nicely and reminds us pleasantly of Australia. Eucalypts are of course now rife in California, in the southwest generally, and practically throughout the world. In Brazil they are used for reforestation and a wood crop. In Tucson there is one next door to us, but we do not want one. The cultivated eucalypts in U.S. strike us as a poor choice among the myriad Australian forms, which vary from what are virtually scruffy weeds (the *mallee* of Australian lingo, for example) to magnificent great trees (the *tuart* of Australia, for example).

On our trip with Dom we stopped at a homestead near Coolup, which had one of the many wonderful Australian place names. The owner, a friend of Dom's, raises cattle (is a *grazier*), but part of his property is a virtual plant and wildlife preserve. Highlights were the great Australian black swans and an open woodland on the floor of which was a carpet of hundreds of orchids of perhaps a dozen species.

On our second day out we accomplished what was for me the most strictly professional purpose of this jaunt: a visit to Mammoth Cave. That cave is neither so mammoth nor so picturesque as the Mammoth Cave in Kentucky, not to mention Carlsbad Cavern in New Mexico, but it has something they lack: it has yielded a large number of fossil mammals and still contains more. Early in the present century collections of these mainly late Pleistocene fossil marsupials were made here by Le Soeuf and a bit later by Glauert, but with poor field records. Now Mike Archer and Duncan Merilees, who had joined us for this visit, were undertaking new, careful collecting for the Western Australia Museum.

South of there we drove on a dirt track through the most enchanting forest imaginable, among noble eucalypts and between banks and great masses of wild flowers of almost all colors of the rainbow. Later, at Cape Leeuwin, a low rocky point of Precambrian granite and gneiss, we were at the southwestern tip of Australia and the eastern end of the Indian Ocean. In the preceeding year we had been at Cape Agulhas in Africa, the western end of that ocean.

Our journey continued to be eventful and I regretfully omit a saga about our chronically ailing car and a succession of nicely helpful local people. I must, however, record an unusual visit with the fauna warden, Arthur Pearce, to a large nature reserve on Two People Bay near the small city of Albany. In 1842, far from here, one John Gilbert had discovered a bird previously unknown to science that was duly recorded as *Atrichornis clamosus* and given the "popular" name *noisy scrub-bird*. From time to time the species was seen here and there around the southwestern angle of the continent, but such sightings stopped in 1889 and naturalists listed the bird as extinct. Then in 1961 an Albany schoolmaster, Harley Webster, whom we met, sprang into well-deserved fame because he saw, heard, and identified a noisy scrub-bird at Two People Bay. At the time of our visit, seven years later, it was known that about sixty pairs were living in a quite restricted area on low Mount Gardner. There probably were none elsewhere. On our visit there we often heard the call, near and far, loud but not really noisy, in fact quite melodious. We did not see the birds, which live in dense cover and rarely fly. The CSIRO had bugged their territories with scattered microphones and lines running to a hut where their cries were automatically and continuously recorded.

My trail has often crossed Darwin's and it did so there. On the voyage of the *Beagle* he spent some time in and around Albany, which already existed then. He detested Australia in general and this part in particular. I love both — I am not Darwin.

From Albany we went north to York and thence back to Perth. Along the way are settlements with wonderful names. For example, Dumbleyung lies between Nippering and Wishbone, which is near Wagin, pronounced as in "wagin' war."

Another highlight was a visit near Maragoonda with a charming couple named Beeck (pronounced "beck") who have a virtual zoo plus nature reserve. There I could become quite closely acquainted with hare wallabies, worlies, and tamars, which are in effect diminutive kangaroos, a nail-tailed wallaby (one of the medium-sized kangaroos), and free mobs of several larger species of kangaroos, as well as a varied selection of brilliantly colored parrots and wrens. It was there that we saw the ridiculous sight, which I have mentioned in another publication, of an emu going full out evidently under the impression that he was a member of a mob of kangaroos whom we had put to flight.

From Perth we were driven to Kalgoorlie. It was on that drive that we passed through a town, Meckering, that was flattened by an earthquake a bit later that day. From Kalgoorlie, by arrangements made by David Ride, we hitched a ride in a tightly packed single-engine, six-seater plane taking supplies to a mineral exploration camp in the Warburton Ranges, a group of hills called "the Warbo" by us old desert rats, far out among the central Australian deserts. There had been a touch of rain recently, and near the Warbo we flew over a dry "lake" that was quite green and was blanketed with purple flowers prettily and appropriately named *Eremophila,* which means "desert lover." We are a bit eremophilous ourselves.

We spent a pleasantly exciting day and night with the hospitable prospectors and then flew on to Alice Springs, having chartered the supply plane that flew us to the Warbo.

Alice Springs is a den of urbanization and iniquity in the midst of one of the broadest, openest spaces on earth. By 1968 it had already acquired fairly extensive tourist traps, from low-class to high-class, but it did still have some of the interesting characteristics of an outpost in a wilderness. The near surroundings are fascinating, quite unlike the flat desert to the south or the hilly desert around the Warburton Ranges. It has a series of wildly undulating hogbacks carved by erosion from great folds in extremely ancient sedimentary rocks. Here and there a watercourse, normally dry, cuts through one of the hogbacks, and we soon had an appropriate picnic at one called Simpson's Gap.

Mike Plane, a vertebrate paleontologist for the government in Canberra, had come out to meet us, and he drove us around, both sightseeing and on more professional business. One of the most interesting bits of sightseeing was a rather long drive, some miles of it down the dry sandy bed of the Finke River, to Palm Valley. That is a sort of box canyon bordered by magnificent cliffs of deep red sandstone with lush vegetation including fairly tall palms (*Livingstonia*) and numerous cycads, relicts in a desert oasis.

On more serious business we drove with Mike Plane and David Ride, who had arrived from Perth via Adelaide, about 105 miles without seeing a house to Alcoota Homestead. The house there, naturally a bit rural but roomy and comfortable with plumbing and home-generated electricity, was not occupied by the owner of the extensive cattle station but by a manager, Ivor Paine, his wife Edna, known as "Ed," and their two youngest children, girls. The girls had a tutor and were receiving lessons and examinations by radio, as on all these remote, isolated stations. The Paines were charming hosts, as for that matter were all the people we ever met in Australia. Thomas Henry Huxley found them so, too. He married a member of a family who entertained him when he visited Australia as a junior officer on H.M.S. *Rattlesnake* in 1847.

Our business at Alcoota was to examine the occurrence and workings of several local deposits of fossil mammals. These had been collected and largely described by Michael Woodburne, a student of R.A. Stirton's at the University of California at Berkeley. Although the Pleistocene mammals of Australia have been known and studied for almost a century and a half now and are fairly well known, pre-Pleistocene mammals are much less known and the finds at Alcoota are among the most important yet made. Our aim at this time was not to renew the work started by Woodburne but just to familiarize ourselves with the site and geology, and we spend a couple of days doing so.

Before leaving the Alice we made an excursion, by light plane, to Ayers Rock, in the far southwest corner of Northern Territory. As almost everyone knows now, Ayers Rock is a tremendous, monolithic, erose dome of bright red sandstone that seems to emerge like the back of some oversized prehistoric monster from the flat and otherwise featureless desert around it. We had noted from the air as we flew from the Warbo to the Alice that Ayers is only one of several such emergent features some miles apart, but it is the largest and best known. Striking enough just in a physiographic way, it has added interest because almost every foot around its perimeter and many places on top figured, and still do figure, in aboriginal myth and religion, and there are fascinating rock paintings in many of the undercuts in its base.

From Alice Springs there were scheduled plane connections to Brisbane by way of Mount Isa, a mining town, and we took them. In Brisbane we met Alan Bartholomai, then recently appointed director of the Queensland Museum, and under his guidance David Ride, Mike Plane, Anne, and I took an excursion inland across the Great Dividing Range to famous fossil mammal localities in the Darling Downs and near the even more curiously named town of Chinchilla. As so often in this account I want to go into details that really are of great interest but that would require a whole library if all were included. I here control myself, but add that we visited Alan again in 1976 and found that he had done wonders with the museum in the meantime.

I also omit details about our visit to Hayman Island, publicized as on the Great Barrier Reef but in fact a high island well shoreward of the reef, about Anne's illness there and hospitalization in the town of Proserpine (pronounced Pross'-er-pine) on the mainland, about another visit to Sydney and environs, and even about our visit to Canberra, one of the most beautiful cities in the world, and my successful lecture there. ("Successful" means that I had a large, attentive audience and that I lived through it.) Just by way of change of pace before moving from Australia to New Zealand I will summarize three items from a Brisbane newspaper:

1. Boy is convicted of killing father and freed on probation.
2. Boy is convicted of hitting (not killing) cat, given three months in jail.
3. Leading headline in full: "BRIDE IS FROM TOOGOOLAWAH." (Are we to infer that not many girls in Toogoolawah bother to get married?)

On second thought, I really must also add the information that in Proserpine there was a shop that sold (exclusively) poisons and paperback books (including murder stories).

So off to New Zealand.

Like much of my travel when I was not traveling for an institution, I went to New Zealand, for a combination of work and pleasure, or I should say work and other pleasures because most of my work is a pleasure for me. The work, to call it that, was to be continuation of my studies of penguins. The other pleasures were to be acquaintance with scenes and people in a country almost completely new to us.

In the last chapter I mentioned my increasing involvement with penguins, both recent and fossil. By 1968 that involvement practically amounted to a new career, a sort of moonlighting job that I have carried on along with a more primary concern with mammals and with less objective problems of theory and the like. In addition to casual visits by strays of a few other species, New Zealand has breeding populations of four species of penguins: little blues, white-flippered, yellow-eyed, and fiordland. On this trip I made a nodding acquaintance with the first three. The fiordland penguins, practically confined to the Tasman Sea side of the South Island, are still among the now very few species of penguins that I have not seen in the wild.

Next to Patagonia, New Zealand is the most prolific source of fossil penguins. Those known from Patagonia are all of nearly or quite the same age, but those known from New Zealand are spread at fairly close intervals all the way from late Eocene to early Recent, a span of some 40 to 45 million years. Most of the specimens are in New Zealand museums, a few in other institutions there, and on this trip I studied almost all found up to 1968. A few found since then or not yet in shape for study then were sent on loan and studied by me in the United States. The very first fossil penguin bone ever found and recognized as such is, however, in the British Museum (Natural History). It was found in 1858 or early 1859 by a Maori whose name was not recorded. He gave it to Dr. W.B.D. Mantell, who sent it on to Thomas Henry Huxley, who published a paper on it and named it *Palaeeudyptes antarcticus*, "The ancient diver from the Antarctic," although in fact it was found farther from the Antarctic than Huxley, in London, was from the Arctic.

On 14 November we flew across the Tasman Sea from Sydney to Christchurch, the metropolis of the South Island of New Zealand. We settled there for a time in a small hotel, more like a first-class English pub. It did indeed have a public bar but also a private bar, comfortable rooms, and a good restaurant. Anne suddenly came down with a serious case of pneumonia, but the hotel personnel were very considerate and helpful, and a good doctor, who found that Anne's lungs "rattled like an old Ford," had her not entirely well but safely able to travel only a few days later.

In the meantime I became acquainted with the excellent Canterbury Museum, Roger Duff, its director and an authority on New Zealand archaeology, and Ron Scarlett, in charge of fossil vertebrates, mostly moas but including some fossil penguins, which I studied. When Anne was a bit better and properly cossetted by the hotel staff, I also made some excursions into the surrounding countryside and in Lyttleton Harbour, which is the port for Christchurch.

Christchurch is an attractive city where, as in the nicest English towns, every house has a brilliant flower garden. The surrounding country is attractive, too, with lush green grass and masses of blooming gorse. But just that touches on why I did not forthwith fall in love with New Zealand and although I now have at least a strong liking for it I still have a slight reservation. The point is that the English settlers of New Zealand seem to have disliked their new home as they found it, and to have done everything possible to Anglicize it. Over much of it they eradicated the native vegetation and replaced it with foreign grasses and trees. Many introduced plants have become uncontrolled weeds, even the yellow lupines, the brier roses, and especially the gorse and broom, rather pretty but here weeds nevertheless. "Acclimitisation societies" also introduced huntable mammals to a land that had none: rabbits, deer, chamois, and tahr from the Old World, wallabies and "possums" (phalangers, not our American opossums) from Australia. These and other introductions became pests, and great effort and expense have been involved in trying to control or eradicate them. We later found that there are reserves and parks with native vegetation and some (nonmammalian) native fauna, but European occupation did widely destroy a desirable and beautiful ecology. It also came close to destroying the native humans, not only by direct attacks on the Maoris but also by arming them with guns. Like all early Polynesians, the Maoris were intensely combative and brutal among themselves. When armed, they did their best to wipe each other out. Virtually none survive on the South Island, and on the North Island only a few persist, almost as museum pieces for tourists.

When Anne could travel, we hired a car and driver to go to Dunedin, farther south on the South Island. We drove across the Canterbury Plains and then up into the mountainous backbone of the island. We crossed the Rangitata, a glacier-fed stream up the valley of which Samuel Butler had his sheep-raising station. He described the mountain region in his satirical fantasy *Erewhon.* We stopped at the Hermitage, a posh government tourist hotel in the midst of the grandest scenery of New Zealand and some of the grandest on earth. It is near the foot of Mount Cook, also near the great Tasman Glacier, and with only slightly lesser peaks, crags, and glaciers in abundance round about. Clouds blowing past Mount Cook gave us only tantalizing glimpses for a time. Butler wrote a century earlier that if you only *think* you have seen Mount Cook, you haven't. Suddenly in a deeply dramatic moment we *knew* we were seeing it, a double summit incredibly high above us. It is only 12,349 feet high, a figure not impressive to a former Coloradoan, but the Hermitage, less than nine miles away, is at an elevation of not over 2,500 feet. There is no place in Colorado where a mountain rises 10,000 feet above an observer less than ten miles distant, and it is that differential that makes for dramatic mountain scenery. Ten-thousand footers are the climbers' acme in New Zealand, and there are twenty-seven of them in this group. It was climbers from here who first reached the top of Mount Everest.

Going back out to the coast down the Waitaki Valley we passed and I inspected the area around the town of Duntroon, which has been the richest source of fossil penguins in New Zealand. Other fossils including the one described by Huxley, have come from near Oamaru, a coastal town we also visited, south of the mouth of the Waitaki. And so on to Dunedin, next to Christchurch the leading metropolis of the

South Island. As Christchurch set out to become more English than England and nearly succeeded, Dunedin set out to become more Scottish than Scotland and also came close. A difference is that unlike older cities everywhere, Dunedin did not start haphazardly, but like Washington, D.C., it was laid out from the start by a geometrical plan that seemed ideal in the nineteenth century but has become unfortunate in the twentieth. In Dunedin the center is an octagon and the older main streets radiate from that. The land is mountainous and wherever you go it seems to be uphill.

We settled in and I was soon happily at work in the Otago Museum, which had the largest single collection of fossil penguins in New Zealand. With hearty cooperation from D. R. R. Foster, director of the museum, and Professor J. D. Campbell of the University of Otago, I soon had a good start on one of my main publications on ancient penguins. With Foster we went out to where we saw a yellow-eyed penguin come ashore; with Campbell, to an important fossil locality.

On 26 November we flew to Wellington, capital of New Zealand, near the southern end of the North Island. There are fossil penguins there in the Dominion Museum and some other interesting material in the Geological Survey in a town with the interesting name of Lower Hutt, which is nearly a suburb of Wellington. When my work there was completed we hired a car and driver and took a long drive to Auckland, a fascinating route which I strongly recommend to visitors but here just designate and do not describe. Our way took us to large Lake Taupo, much frequented by fishermen; nearby Wairakei, a geothermal plant in full swing and of much interest in the looming energy crisis; to Rotorua, which has the twin attractions of one of the three true geyser districts in the world and a quite touristy Maori settlement (Whaksarewarewa; our Maori guide there was named, or at least called, "Bubbles"); through Hamilton, largest inland city in New Zealand and Pacific headquarters of the Church of Jesus Christ of Latter Day Saints (Mormons); and Auckland. There we boarded a ship, the (then) Matson Line's *Monterey*, which took us back to Sydney for a few days and then across the Pacific to home.

Notes

§ I have already mentioned that Nadi is pronounced (more or less) "nahndy." The Fijians had no native writing and so a system using the English alphabet was devised for them by the uneaten missionaries.

In that system, still in general use, *b* is pronounced "mb," *c* is pronounced "th" as in "this," not as in "thin," *d* is pronounced "nd," *g* is pronounced "ng" as in "singer," not as in "finger" and *q* is usually but not always pronounced "ng" as in "finger," not as in "singer." The letters *f, h, j, p, x,* and *z* are not used. The system looks odd, offhand, and causes tourists to mispronounce most Fijian toponyms and street names, but it does make sense in terms of strictly Fijian phonetics.

§   Ediacara is accented on the third syllable: edia'-cara.

§   There is no "aboriginal language" in Australia. There are dozens of them, a veritable Babel with only a few people speaking any one. An example is provided by just a few of the Western Australian abo names for the black swan, *Cygnus atratus* to white ornithologists, to abos: kulla-eedoo, kurillthu, guroyl, mallee, kooljak, weeler, and on and on. The abo Ballaroke family of the Bibbulman tribe are descendants of black swans who turned into humans.

§   The Warburton Ranges are named for Major (later Colonel) Egerton Warburton, who traversed this previously unknown region in 1873. Alice Springs was originally named Stuart and it has been said that the name was changed in honor of Princess Alice, daughter of Queen Victoria. The local and more likely story, however, is that the name was first applied to a waterhole, near but not at Stuart, named by a surveyor, W. W. Mills, for Alice Todd, the wife of Mills's boss C. H. Todd.
From the fact that his name is applied to a large desert and to some other landscape features one would think that one Simpson (not related to me in any case) must have been of some note in central Australia, but in considerable reading of Australian history and exploration I have failed to find his name. In Alice Springs, I do not know on what authority, I was told that the relevant Simpson had never been in central Australia but was a manufacturer who financed some exploration there.

§   There really is a town named Toogoolawah, from which the publicized bride came. If you care to, you can find it at latitude 27°5' south, longitude 152°19' east.

§   New Zealand was formerly divided into administrative districts that are not now political units but the names of which are still in universal use. Christchurch is in Canterbury and Dunedin in Otago, whence the names of their museums. These museums and some others in New Zealand receive a certain percentage of the national taxes as well as support from various other sources. I do not know any country better provided with excellent natural history museums in proportion to population. The War Memorial Museum in Auckland must also be mentioned.

§   The true geyser districts in addition to Rotorua are in Iceland, the original one, and Yellowstone Park.

References

Barrow, T. 1964. *The decorative arts of the New Zealand Maori.* Wellington, Auckland, and Sydney: A. H. and A. W. Read. Includes all the arts still practiced and many no longer so.

Connor, H. E., ed. 1966. *Mount Cook National Park.* 4th ed. Christchurch: Mount Cook National Park Board. An unusually good tourist guide.

Glaessner, M. F. 1961. Pre-Cambrian animals. *Scientific American,* March 1961, cover and p. 72–78. A popular article on the Ediacara fauna.

McCarthy, F. D. 1966. *Australian aboriginal decorative art.* Sydney: Australian Museum. A brief but good introduction to a subject that has a very extensive literature.

Milner, G. B. 1967. *Fijian grammar.* Suva: Fijian Government Printing Department. A good introduction to one of the world's strangest languages.

Morecombe, M. K. 1968. *Australia's Western Wildflowers.* Perth: Landfall Press. A stunning book with great color photos not only of the finest wildflower display on earth but also of the unique mammals, birds, and insects that derive food from them.

Mountford, C. P. 1965. *Ayres rock: its people, their beliefs, and their art.* Honolulu: East-West Center Press. The best account I know.

Serventy, D. L., and H. M. Whitell. 1967. *Birds of Western Australia.* Perth: Lamb Publications. A standard work; the senior author introduced us to his subjects.

Serventy, V. 1967. *Nature walkabout.* Sydney, Wellington, and Auckland: A. H. and A. W. Read. Perceptive account of a long trek around Australia, the track of which we crossed several times. The author is Dom Serventy's brother.

Simpson, G. G. 1971. A review of the pre-Pliocene penguins of New Zealand. *Bulletin of the American Museum of Natural History* 144: p. 319–78. This is primarily a result of my visit to New Zealand in 1968 and is the most substantial of my several studies of the fossil penguins of that country.

White, H. L., ed. 1954. *Canberra: a nation's capital.* London: Angus and Robertson. As I have not taken space to describe this admirable city, I give this reference to a full treatment.

22

Journeys South

Penguins are habit forming. — G. G. Simpson

My friend and colleague Larry Gould considers it infra dig to voyage to the Antarctic as a tourist, and so it would be for him. He is an old Antarctic hand who has traveled there the hard way and who even now, in just appreciation of his achievements, is periodically flown to the Antarctic under the most official of auspices. For me, however, with my infirmities, my absence of relevant achievement, and my lack of official connections, the only way to get there seemed to be as a tourist. So I have gone there, with Anne, three times on the Lindblad *Explorer*. Larry still speaks to us.

The *Explorer* stands well up in our ratings of the many ships on which we have sailed. She is small, taking fewer than a hundred passengers and no cargo, and correspondingly lacks some of the advantages as well as disadvantages of larger ships. We are glad to be without some of those and can manage without any of them. The *Explorer* has her own advantages, such as a lecture room and the usual presence of well-informed naturalists rather than supposed entertainers. With inevitable exceptions the passengers are usually mature, educated people with a real interest in places and in nature. Many are birders, but some of our best friends are birders. There are no cruise bums, although most passengers have taken more than one cruise on this ship. We have taken six and hope to take another. The first three of the six were to Antarctica.

Our first attempt did begin badly. We were booked on the *Explorer's* first Antarctic cruise in 1970, but the night before we were to leave by plane for Argentina the cruise was called off because of an accident on the ship. We planned anew to visit colleagues in Argentina and take a later cruise, but Rosendo Pascual, with whom we were to visit Patagonia, had an accident that put him in the hospital between life and death all the time we were in Argentina, and longer; happily he did finally recover fully but the effects were long-lasting. Our plane flight was indefinitely delayed, and a substitute landed us at Azeiza, the international airport for Buenos Aires, in the middle of the night of 26-27 January 1970, without transport for the many miles into the city. Problems went on and on, but all were more or less satisfactorily solved.

We drove down to La Plata, capital of Buenos Aires Province and within rather long commuting distance of the city of Buenos Aires, which is not in the province of that name but in a federal district. There we were cordially received by colleagues and friends, notably Rosendo's charming wife Nelly. (Nelly is a not uncommon name among upper-class Argentinians despite its decidedly non-Spanish origin; we also have a very Argentine friend whose name is Mary, not María.) I visited Rosendo in the hospital and he reacted with apparent pleasure to my presence, but he never thereafter remembered my visit. I put in a good bit of work in the Museo de La Plata, among other things on studies of two extraordinary fossil marsupials, which I published after my return to the United States. This was done a few months before my connection with the M.C.Z. was terminated, and those were my last two publications as from that institution (these jointly with the University of Arizona).

We flew down to Mar del Plata and had a pleasant visit with the Scaglias and at the Museo Municipal de Ciencias Naturales de Mar del Plata, now in a fine new building, as I mentioned before. I went over some of the collections and selected some specimens to be studied later in Tucson on loan. Thence we returned to Buenos Aires and on 7 February flew from there to Trelew in Patagonia, which had changed considerably since my first visit (mentioned in chapter 7). It still had the same hotel, but to my amazement now with some private bathrooms, of a sort. There were some Argentine geologists working near there and we enjoyed their company. A vertebrate paleontologist, Rodolfo Casamiquela, had recently moved to the neighboring province of Río Negro, and he drove down to meet us. With him we drove out the Peninsula Valdés and enjoyed seeing large numbers of southern sea lions (*lobos* to Argentinians) and southern elephant seals and just one penguin. Despite that pleasant glimpse of wild life, Patagonia had changed enormously in the forty years since I first saw it, and in my opinion not for the better. I wrote at the time; "It is incredible and terrible that we drove some twelve hours and some hundreds of kilometers and did not see *a single* guanaco or rhea. Patagonia is still bleak, windy, and wide, but it is no longer the wild, free end of the earth."

Another flight took us to Comodoro Rivadavia. Here, too, the hotel where we used to stay when in town in the 1930s was still standing, but it was almost the only relic from that period. There were two ultra-modern, high-rise hotels, and we stayed at one of them but did not patronize its nightclub. (Night life there in the '30s involved only the licensed brothels.) We drove out to Colonia Sarmiento, or rather to Sarmiento, for *Colonia* had been deleted from the name, in order to visit my old and valued field assistant and friend Justino Hernández. The drive itself was rather sad because we drove where animals used to swarm — several kinds always in sight — and now we did not see even one. The visit with Justino, however, was delightful and hilarious — two grandfathers recalling their rugged and spirited days together when they were young. We also met for the first time and liked very much Justino's wife, a charming Lithuanian woman. (You may recall that Justino is half Lithuanian and half Araucanian Indian.) We were joined by Justino's uncle, Don Casimiro Szlapelis, well into his seventies and flying his own plane after wrecking two and walking away from them. After a touch of wine Justino and I tried some of the old songs, such as "Lo' Mendocino' " and "Mañana es Domingo," but inexplicably something had happened to our voices and our memories.

Leaving Comodoro for Punta Arenas we changed planes and lunched in Río Gallegos. Anne wants me to include the menu of our table d'hôte lunch at the Gran Hotel de Paris:

1. Slices of cold pot roast dressed with chopped egg, pimento, chopped parsley, and vinegar
2. Mutton broth with pastina
3. Tripe cooked with salt pork, potatoes, carrots, and a little onion
4. Fried beefsteak with tomato salad
5. Chocolate pudding or stewed prunes
6. Coffee

— with plentiful Argentine wine and bottled spring water.

It is significant of what is happening to Patagonia and to the world that there was a souvenir shop in this still really very remote spot — at 51°35′ south latitude few people have ever been so far south, and we had never been until then. They had a pitcher in the form of a penguin with golden beak, eyes, and wings, with flowers that may or may not be roses on its belly, and with RECUERDO DE RIO GALLEGOS crudely printed above them. This was too ridiculous to resist, and it is before me as I now write about it.

In Punta Arenas, a fair-sized and interesting city on the Strait of Magellan in Chile, we met up with, or rather were joined by, our tour, at the Hotel Cabo de Hornos, which means "Cape Horn" in Spanish (but see note following) although Punta Arenas is far from Cape Horn. Gunther in his *Inside South America* speaks well of this hotel and particularly of its martini cocktails. We liked the hotel well enough but found the martinis horrible. Perhaps they had changed bartenders, or perhaps Gunther understood politics more and drinks less than we.

After some backing and filling we took off on what was the new ship *Explorer*'s actual second but scheduled third tour. From the Strait of Magellan we went out briefly into the Pacific and then eastward among islands and into the Beagle Canal, which is named for the ship on which Charles Darwin went around the world and which separates the large island of Tierra del Fuego from smaller but still considerable islands south of it. The scenery is magnificent — perhaps I should have a rubber stamp made of that, for from here on all the scenery was magnificent and no one who sees it can help saying so over and over again. On just this stretch the jagged Darwin Range, north of the Beagle Canal, is especially magnificent, with many great glaciers, some to water's edge and others with cataracts below them. Ushuaia, east of the Darwin Range and also on the north side of the canal (a natural water passage, not a canal in the more usual English sense), is the southernmost town in the world. On this occasion we did not stop there. Tierra del Fuego is part Chilean and part Argentine, with Ushuaia in the Argentine part. We had a Chilean pilot who would have been arrested if caught practicing his profession in Argentine waters.

We went out into the Drake Channel or Passage, the waters between the near-coastal islands of South American and those of Antarctica. Ever since sailors started rounding the Horn, Drake among the earliest, this passage has been known as one of the roughest on earth. It has lived up to that reputation on our several crossings of it. Anne never has been and I was no longer subject to seasickness and the *Explorer* is a relatively steady small ship, but she did roll madly for a time. That does not interfere with my eating or drinking but it does with my walking, so on such occasions I tend to stay put in our cabin.

On the afternoon of 18 February we sighted a cloud that turned out to be the snow-capped peak of Smith Island. We did not stop there but turned to port, eastward, and came to our first Antarctic anchorage, inside Deception Island in the South Shetlands. It seems odd to write of anchoring inside an island, but ships do so there. Deception was once an immense volcanic complex; when it collapsed, it formed a central basin, or in geological terms, a *caldera*. There is one break in the rim, now called Neptune's Bellows because of the winds that tear through it, and through this ships can enter to the flooded, almost completely landlocked waters in the caldera. Volcanic activity is still very much alive there, and two research stations were completely destroyed by ash falls in the few years before our visit. Almost miraculously no lives were lost. In places the water is warm or even hot, although those who tried swimming found it deceptive — it could be uncomfortably hot at the surface but cold and Antarctic just below. We were run around the caldera in Zodiacs, the inflated rubber landing boats with outboard motors used from the *Explorer* for shallow-water landings and exploration. The landscape was both lunar — drab lava and ash, no vegetation at all, mountainous — and Antarctic in another sense — snow, ice, overcast, bitter cold even to us in long underwear, ski pants, sweaters, bright red parkas, balaklava helmets, and boots. Our introduction to the Antarctic was one of the most unusual but not the most attractive of our days there.

We proceeded eastward to the island variously called King George I or Veinticinco de Mayo, among other names. The United States has made no territorial claims on Antarctica or the adjacent islands, but there are conflicting claims over much of this vast region, with the result that many places have several names from different languages and nations. Here British and Argentine claims overlap, but the two research stations are Chilean, the Eduardo Frei, and Russian, the Bellingshausen — The Russians call this "Bellinggauzen" and they stamped our passports, more as a souvenir than for any official reason, with the date 1969 although in fact we visited them on 19 February 1970. The two stations are only a few hundred yards apart, separated by a small stream, and the Argentinian naturalist on our ship explained that *malas lenguas* (literally "bad tongues," i.e., malicious gossips) say that the U.S.A. had the Chilean station installed to keep an eye on the Russians. Anne quite properly told him that he was himself a *mala lengua* to invent or repeat such a ridiculous statement. We were well received at both stations but could not make out what, if anything, they were accomplishing.

We made another landing across the bay and there visited a fairly large penguin rookery with both chinstraps and gentoos. I had seen magellanics in Patagonia, black-footed penguins in South Africa, and three more species in New Zealand, but these two more southerly species were new to me on this trip. There is a large rookery of chinstraps on Deception Island but I did not get a really good look at this species until we reached King George I Island. A couple of years later I did also get in good chinstrap watching on Deception. It is quite possible that yóu have seen them there, too, though more remotely, as they figured prominently, if not adequately identified as to species and locality, in a TV show by Jacques Cousteau. (Cousteau's party was there when we were two years later.) The scene on King George I: "Antarctic as all get-out. Black cliffs. Green slopes. Gray sky. White snow banks with bands of penguins busily waddling up and down them. An inlet with a big glacier to the sea, calving icebergs."

The next day we went into Gerlache Strait and the Canal Argentino, saw the mainland of Antarctica for the first time, and set foot on it at Gonzalez Videla, an abandoned Chilean research station. This had been taken over by a large, unruly, clamorous, and odoriferous mob of gentoo penguins and their chicks. There were also many sheathbills about, white and clean looking but in fact rather filthy scavengers. They are the only Antarctic birds without webbed feet. In the canal ("canal" in the Spanish sense, a natural channel) many groups of penguins were playing follow-the-leader, porpoising up into the air and then down below the surface, and gathering krill, small crustaceans (euphausiids), later to be regurgitated for their demanding young on their pebble-pile nests ashore. In this region, especially in well-named Paradise Bay, kelp gulls and skuas were also common, the latter occasional predators on young or injured penguins. Leopard seals, the most dangerous predators on penguins, were also about, as were crabeater seals, which do not eat crabs but share the krill with penguins and with whales. Weddell seals are abundant everywhere in the Antarctic. Among gray stones a somnolent Weddell seal looks so much like another stone that one of our fellow passengers stepped on one before he discovered that it was alive. James Weddell (1787–1834) was a professional killer, not admirer, of seals, and it is ironic that seals (and also an Antarctic sea) should be named after him, but he did make this species known to the British zoologists.

The scenery here and on southward along the west coast of the Antarctic Peninsula is magnificent beyond comparison. Although I must again qualify a superlative because I have not seen the Himalayas, this is to me the grandest scenery on earth. I have now also traveled extensively in Alaska, northern Canada, and Greenland, and there is grand scenery there, but nothing to match that of the Antarctic Peninsula. That goes for the Rockies, the Alps, and the Andes, too, all of which I know fairly well. To be sure, one of the more naïve of our generally sophisticated fellow passengers "just couldn't get over the fact that Antarctica is exactly like Switzerland," which is like finding the Sahara exactly like Kansas because both are flat and rather dry.

We were generally traveling between the peninsular mainland and a series of islands. Black mountains rise precipitously from the water, sculptured into a million jagged forms. The rocks, exposed on the many nearly vertical surfaces, are mostly dark but the tops of the mountains and slopes less than vertical are snow-covered. Every valley has a living glacier, tumbling to the sea and breaking off, leaving ice walls of most delicate blue. The sky is pale aquamarine at the horizon, pale azure at the zenith. The sea is steely where deep, pale nile-green where shallow or over a ledge of ice. Even in the narrow channels there are small bergs and floes. On many of them seals were enjoying the balmy weather — it was summer and the air temperature was generally a little above freezing. The sun does set, but the short nights are not fully dark.

Beyond the narrow Lemaire Channel are broad, still waters between the jagged peaks, the sea sprinkled all over with floating ice from chunks small enough for a highball to bergs half a mile or more across. The usual bergs of moderate size, say twenty to a hundred yards across, are sculptured in myriads of fantastic shapes: smooth, jagged, round, square, flat-topped, and pinnacled, with windows, tunnels, and every variety of surrealist sculpture. Almost all of the lower and more or less flat bergs have groups of crabeater seals on board, just lying around taking it easy on the nice warm ice.

As we cruised around there a small ice cake drifted by with two apparently somewhat bewildered penguins on board — adélies, the little gentlemen in dress suits that most people think of as *the* penguin. The first we had seen — though we were to see hundreds of thousands of them before we finally left the Antarctic. Although it is a common notion that penguins are all antarctic birds, in fact of the sixteen living species only the adélies and the emperors are strictly antarctic and even they occasionally stray into lower latitudes.

The night of 21–22 February we stalled in heavy pack ice, but the captain managed to extricate the vessel and get it out into rough open seas. At about 2:45 A.M. we were awakened by the news that we were crossing the Antarctic Circle and many of us, *en déshabillé,* gathered in the lounge-bar and toasted the officers, guides, each other, and the world in champagne. On other gala occasions one thing we did not like about the Scandinavian management was that toasting was done in something called (I believe — I do not find it in Norwegian or Swedish dictionaries) *glögg,* which is no more attractive than its name.

On a small island called Thorgeson or Torgeson quite near Anvers, on which is the American Palmer Station, we visited our first large adélie rookery, replete with youngsters in their second downy plumage. In the downy stages the plumage is not waterproof, so the chicks cannot yet go to sea or feed themselves and both parents are kept on the run feeding them. The pack ice drifted in and we had trouble getting back to the ships in the rubber Zodiacs, not exactly ice-breakers. Ours went ahead as we fended off the ice as best we could with oars, another Zodiac got a nearly free ride in our wake, and behind it three seals took advantage of our efforts and followed suit. We warned that they were probably leopard seals waiting for us to fall overboard, but in fact they were inoffensive crabeaters.

After further adventures that need not be detailed we went back across the Drake Channel, past Cape Horn, into the Beagle Canal, and this time stopped at Ushuaia. We took an excursion in a fleet of taxis into a national park and to the shore of Lago Rocca, a large lake straddling the Argentine-Chilean border. This and other waters of the region are stocked with trout and salmon to lure sportsmen. My ideas of conservation and even of sport would prohibit importing foreign fish or animals to be killed for sport or of plants to replace native ones. Incidentally, that national park teems with European rabbits, a more obviously unwise importation. We did pass through some handsome forests of the southern beech, *Nothofagus,* which interested me especially as we had also seen native *Nothofagus* in Australia and this is evidence that the two continents were once connected via Antarctica.

By ship back to Punta Arenas and by air back to Tucson, with stopovers in Santiago — which then, at least, had one of the two worst airports in the world (the other was the domestic airport in Buenos Aires) — Lima, and Mexico City. I can't resist telling just one incident during a stopover. I was going through the rather scanty exhibits of fossils in the Museo Nacional de Historia Natural in Santiago when I was taken aback to come upon a picture of myself among them.

For our next visit to Antarctica we joined the tour in Los Angeles on 10 January 1971 and flew to Hobart with a stopover in and about Sydney. We found Hobart, capital of Tasmania, a charming small city. There we boarded the *Explorer*. The next few days we were in exceptionally rough seas, as bad as those of the Drake Channel, but we plowed southward to Macquarie Island, where by remarkably good fortune we were able to land. This highly isolated subantarctic island has much bad weather and heavy surf and has no harbor, so that for weeks at a time it may be impossible to land there. It is the only place where royal penguins regularly breed, and there are also king, gentoo, and rockhopper penguins there. We had seen lots of gentoos on our first Antarctic trip, but the other three were new to me. We did see a great many more rockhoppers on our third Antarctic trip, but this is the only place where we saw royals and we saw only one king elsewhere. The royals are handsome with their white faces, thick pink beaks, and jaunty yellow crests, but the kings are the most beautiful of penguins and to me, at least, among the most beautiful of all birds. They have blue-gray backs and flippers, black heads, white fronts, and bright yellow sometimes almost orange throats and patches back of the eyes. (The frontispiece of my book on penguins is a color photograph of three of them that I took that day on Macquarie Island.)

There are no trees on Macquarie but it is vividly green, almost covered with low, emerald vegetation. It is the southernmost *green* island in the world, latitude 54°29' south. Every place, everything, and everyone can be unique in *some* way, as Roger Peterson once remarked about a friend believed to be the only person in the world to have performed mouth-to-beak resuscitation on a rockhopper penguin. Also extraordinary on Macquarie are the elephant seals, hundreds of them — in some places the ground is literally paved with them. They seem to spend all their time lying around, and I don't see how they make a living. They are so inert you might fancy them dead, but if you come close they do raise their heads, open their bright pink mouths, and roar to scare you away.

From Macquarie we were again several days on the open sea and on 21 January crossed the Antarctic Circle with no difficulty and with salutes in glögg rather than champagne. Here we began to see icebergs and whales, mostly the medium-sized seis. These waters once teemed with whales, but they have been killed off to near extinction, an activity that civilized countries have stopped but the Russians and Japanese still carry on. The next day we were in pack ice and saw land for the first time since Macquarie: Coulman Island to port and Victoria Land, part of the continental mainland, to starboard. We couldn't get through the ice there, so had to backtrack and continue outside Coulman. As we bore away southward there was an emperor penguin, largest of living penguins, on an ice pan and we got a good look. This was rare good fortune because at this time of year, midsummer, the emperors are all out of their rookeries and scattered at sea.

Once out of the pack ice we hit a tremendous head wind, force 8 or worse, and wallowed along until late on the afternoon of the twenty-fourth we came to the Ross ice shelf, a grand sight, a continuous but jagged wall of white ice rising over a hundred feet above the water. Many adélie penguins were swimming along the front, porpoising in the peculiar and delightful way they have when going all out. Mount Terror was in sight, a quiescent but perhaps not extinct volcano. It is not named for the effect it causes but for one of Ross's two ships. There is also a Mount Erebus, for the other one. Erebus, which we saw later, is an active volcano. Sir James Clark Ross (1800–1862), unlike the sealer Weddell, was a genuine scientific explorer both in the Antarctic and in the Arctic, and I do not begrudge him the Ross Sea, Ross Ice Shelf, Ross Island, and Ross seal. We skirted the north side of Ross Island, the land mostly deep in ice but with one long clear area on which was a rookery of *several hundred thousand* adélie penguins. Next morning when we woke up we were moored at McMurdo.

McMurdo is the main American base in Antarctica and at the time of our visit it had summer accommodations for about a thousand men, a few doing some research but mostly naval support, and even a few women. Some, but many fewer, also wintered there. After my return I reported through proper channels a poor opinion of McMurdo and I am informed that it has been reorganized and improved since then. We also visited nearby Scott Base, the principal New Zealand base in Antarctica, then with about fifty men, of whom about a dozen over-wintered. Their base was neater and better managed than McMurdo and the personnel seemed more competent. It must be said, however, that the much smaller size makes for neatness and efficiency and also that Scott was receiving some support from McMurdo.

They still had sledge dogs at Scott Base and did use them — dogs don't need spare parts and they run on local fuel: seal meat — but I believe they were kept more for sentiment. They were staked out on the ice a mile or so from the base and lived in the open there the year around. (I have since learned that kennels have been provided.) This is a relatively warm spot for Antarctica and the winter temperature rarely goes below minus 50° Fahrenheit. The dogs were delighted to have company and begged to be petted. They love people but hate each other and have to be kept tethered out of reach of each other.

We got away from McMurdo on the twenty-seventh and went down the west side of Ross Island to Cape Evans, where there is one of the two Scott huts (the other is at McMurdo), relics of his ill-fated expedition, and to Cape Royds, where there are adélie penguins and there was one chinstrap, south of the previously recorded range of that species. We turned north again and ran down the coast of Victoria Land with a stop, in bad weather, at Hallett, a joint U.S.-New Zealand base since abandoned. This coast, unlike the area around McMurdo, is extremely picturesque and we greatly enjoyed the trip along it, but it is picturesque in the same way as and in less degree than the Antarctic Peninsula so I will not take space here for details.

After two rough days at sea we landed at Campbell Island. Sheep and cattle farming was carried on there from 1895 to 1931. At latitude 52°30′ south the island is only marginally subantarctic, but its climate and remote situation make animal husbandry unprofitable. The settlers abandoned it, leaving a shocking legacy of feral animals and introduced weeds. It is now a New Zealand government reserve with a research station on Perseverance Harbour, a picturesque fiord that extends more than half way across the island. We entered there and visited the station. Near the entrance is a large rookery of rockhopper penguins, and there we also saw one miserable king penguin, well outside its breeding range, that had hauled ashore to molt. There are innumerable other sea birds, notably royal albatrosses nesting high above the station and floating everywhere in the sky with probably the greatest wing span of any living birds. Other albatrosses — sooty, wandering, mollymawks — cape pigeons (not pigeons but spotted petrels), skuas, gulls, ducks, cormorants ("Campbell Island shags"), and others make this a sea-bird watcher's or birder's paradise. Although the island is largely volcanic there are sedimentary rocks as well and in a limestone many chert nodules that some passengers took for gem agates and collected in quantity. They were annoyed and unbelieving when I assured them that these were not agates and would not be a commercial proposition if they were. However, they were good souvenirs — I have one here now.

Next day we were at the Auckland Islands (which are very remote from the city of Auckland) and landed on the northernmost and flattest, Enderby, where there were hundreds of southern or Hooker's sea lions, past pupping and now copulating, the males dangerously aggressive toward intruders, even human ones. Even more interesting were many yellow-eyed penguins, not in big rookeries because they are not particularly colonial or social, but in small mobs of a dozen here and six there.

Back to New Zealand, South Island, at Port Bluff and by train from nearby Invercargill to Christchurch. We had planned to go on to New Guinea, but Anne went into heart failure and spent some time well tended in an excellent, no-frills but all-care hospital. When she could travel, with wheelchairs and rest, we flew back to Tucson.

We took off from Tucson on our third and so far, at least, last Antarctic journey on 6 December 1972. We flew to Buenos Aires with Martha Eastlake, my sister, and after a short stay there to Mendoza, our favorite city in Argentina. We enjoyed visits with our old friends the Minoprios. My even older friend Carlos Rusconi, whom I had met in Buenos Aires in 1930 and who later was director, and virtually the whole staff, of the small museum in Mendoza, had died since our last visit. Then by air for our first visit to Tucumán, where we were put up in a comfortable hotel as guests of the Instituto Miguel Lillo, an excellent research institution devoted to natural history. There is an able vertebrate paleontologist there, José Bonaparte, and a likewise able herpetologist, Raymond Laurent, of Belgian origin and having been for some time in the Belgian Congo. He had been at the M.C.Z. part of the time while I was there.

At the Lillo, Bonaparte generously showed me his pets, mostly Triassic reptiles and including one so near the arbitrary line between reptiles and mammals as some of the former evolved into the latter that it could about equally well be put into either class. I also went over many of the much later fossil mammals in the Lillo collection and found several of special interest to me, which I included in subsequent publications on fossil marsupials. In the library an attractive librarian in a T-shirt emblazoned TEXAS TECH showed us bibliophiles' treasures of the venerable institution and presented us with some beautiful large folio loose plates from an early flora of Argentina. A series of them, suitably framed, now graces our living room in Tucson. I owe it to the librarian to add that at a meeting later that day she was dressed very elegantly indeed.

We were hospitably, indeed royally, treated by numerous old friends and new acquaintances in Tucumán. The only return I could make was to hold a round table, so-called, which turned out to be my fielding hard questions in Castellano from an interested and exigent audience of a hundred people or more.

Back in Buenos Aires we joined the *Explorer* tour on 22 December, went by train to Mar del Plata, and there boarded the ship, which sailed that night. After two days at sea we docked on Christmas morning in Port Stanley, the only town on the Falkland Islands. My sister Marty had read that this remote outpost looks like an English country village, and she was a bit indignant because to her it did not. I would say that it looks like an English village not, to be sure, in England but in latitude 51°45′ south. It was summer there and the most striking thing as we wandered around town was the display of masses of flowers in front of every cottage, mostly lupines, spiky and rather too stiff for my taste but a brave display. Wherever the English go, their descendants plant flowers and harvest Brussels sprouts, belief in the edibility of which is one of the peculiarities of that staunch breed. Most of the cottages also have conservatories or hothouses full of potted plants. Heating is by fireplaces stoked with peat, enormously present on these islands, and the burning peat gives the whole town a distinctive, rather pleasant odor.

As this was Christmas almost everyone was holed up at home, although one small shop finally did open to take advantage of the only tourist visit of the year. We did not take advantage of that, but we did regret the fact that the town's one bar was closed because we thus did not meet the barmaid who is immortalized by the name Eunice's Teats given to massive, shapely twin peaks on the Antarctic Peninsula.

The Falklands had no aboriginal inhabitants and were probably first sighted by John Davis, an English navigator, in August 1592. The first occupation was in February 1764 by settlers from St. Malo, France, and French descendants going into exile from Acadia, Nova Scotia. The subsequent history is both complex and disputed. For details see the book by Strange cited at the end of this chapter. Here it suffices to say that after rather sporadic occupations by the French, English, and Spanish with intervals of lawless, sparse population by whalers and sealers, the British established what has so far been complete control precariously in 1832. Argentina still claims to have inherited legal ownership from Spain and Argentinians always call the islands Malvinas, which is the Spanish equivalent of the French *Malouines,* given to the islands in honor of their patron saint by the French from St. Malo. They, not the Spanish, were in fact the first settlers.

Recently an Argentinian expedition "explored" the "Malvinas" and reported their observations in a publication that conveys hardly a hint that they were essentially guests of the British there and that they added nothing to what had already been made known, mostly by the British. I am devoted to both the Argentinians and the British, but I characterize this episode as I must judge it.

The Falklands consist of two large islands, East and West Falkland, and almost innumerable lesser islands. Christmas night and the next morning we went through heavy seas around the northern coasts of the main islands and anchored in the harbor on Westpoint, a small island off a northwestern peninsula of West Falkland. There was a settlement there, only five or six houses. Some of us, including Marty and me, struggled up the steep slope to go over a divide and were pretty well winded when offered a lift in a four-wheel-drive vehicle driven by a nice young girl from the settlement. Drive us she did on that trackless land, full out, jittering, bouncing, and leaping over peat bogs and rocks, and unloaded us at a fence on the divide. When we thanked her, she assured us that the pleasure was all hers: she had never driven a car before and was delighted to have had the chance. With trembling legs we made our way on foot down to a hollow with great piles of rocks all covered with rockhopper penguins, shouting their heads off at each other and at us. The young were just hatching. Black-browed albatrosses were hanging about in the penguin rookery, for reasons best known to themselves.

We went back from the divide by different but equally reckless transportation. A truck was there, apparently abandoned, and a local woman asked whether any of us could drive it. One of our female companions volunteered and was given a quick lesson:

"What's this thing sticking up?"
"The gear shift."
"You mean it doesn't shift gears by itself?"
"No. You have to do it. Here is first and here is second. You won't need others."
"How do I know where to go?"
"Follow the tractor."

We survived.

Later, at Carcass Island, we took a Zodiac to look for elephant seals, which we did not find. That did not matter to us because on previous trips we had had almost a surfeit of elephant seals. It did result in probably the loveliest marine experience I have ever had. A group of five or more of the beautiful pied or striped black and white Peale's porpoises (a species of *Lagenorhynchus*) came, greeted us, and sported around our boat for a long time. They would surface and blow just beside us, cut dashingly across our bow, come directly below us where we could see them in the crystal-clear water among the kelp, and then swim right under the boat. They were enchanting. There were penguins about, too, here all magellanics, the same species as on the Patagonian coast.

Under the guidance of Ian Strange, an islander who probably knows the Falklands better than anyone else, we went on southwest to New Island. In one place there we were entertained by gentoo penguins, a relatively small group and all prebreeding, what Roger Peterson called "teen-agers" or "unemployed." They were preparing for later responsibilities of nest building and parenthood by stealing pebble-sized lumps of mud from each other. From their behavior they could also legitimately be called hoodlums. After a trip by Zodiac and a long, for me very difficult, overland hike we also visited a rockhopper rookery with, conservatively estimated, one hundred thousand occupants. A few couples of macaroni penguins had also taken up residence there and this delighted me because I had not seen that species (*Eudyptes chrysolophus*, "the golden-crested expert diver") before.

From New Island we went along the picturesque south coast of West Falkland, checked out officially at Fox Bay, and took off for Antarctica. Although far from identical, this journey near and along the Antarctic Peninsula was similar to the one Anne and I had taken in 1970. We could take it at least a dozen times more without tiring of it, but I will avoid repetition here.

Our first landing was at Potters Cove on King George I island. There was a small mixed band of unemployed penguins, gentoos, adélies, and chinstraps, perhaps fifty in all, inveterate people watchers, who followed us about. Here I took my favorite among thousands of pictures that I have taken — I used to be a rabid photographer but except for pictures needed as records or for publication I no longer care much about them. This one has an ice-covered low island in the background, deep blue open water nearer, then the beach with a pile of elephant seals mostly somnolent but three rearing and roaring at each other, and in the foreground one adélie, stepping out in the self-important way they have, and evidently late for an important engagement. It is reproduced in color in my book on penguins.

Ice conditions being favorable, this time we got out to the very tip of the Antarctic Peninsula and had a look into the Weddell Sea. Most of the splendid journey down the west side of the peninsula was happily similar to our first one. There was some pack ice but not enough to be troublesome, and the weather, although of course cool, was fine. The Antarctic Circle was an old story to this ship by now and celebration was in glögg, not champagne, inspiring Marty to one of the world's shortest poems:

> Glögg —
> Ugh!

Under continuing favorable conditions we went considerably further south than in 1970, reaching the British station on Adelaide Island.

Since our first visit there I had become acquainted with the Palmer Station scientific leader and other Palmer Station scientists while I was lecturing at the University of California at Davis, and we had a particularly hospitable welcome at this station.

Our shipmates were still almost all pleasant people who were keenly and intelligently interested in where we were and what we were seeing, but after a number of Antarctic tours there were now a few nice old ladies who were just along for the ride and not quite sure where we were or very interested in what we were doing. Barbara Peterson, then still married to Roger, and we called them "The Biddies" and collected sayings by them. I give only two of many examples:

1. Biddy to Anne, while we were at sea, without reference to a map or other antecedent: "Are we here?"
Anne: Yes.
Biddy trots off perfectly satisfied.

2. On return to Ushuaia from trip to Lago Rocca. Naturalist guide, at a scenic stop: "You can get out here for five minutes if you want to take pictures."
Biddy: Can I get out if I don't take a picture?

We had a pleasant time in Ushuaia, where I was delighted to meet Mrs. Goodall, an American who had married into the Bridges family (see reference following) and moved to Tierra del Fuego to live.

Then back by stages to Tucson.

Notes

§ The names *Cape Horn* in English and *Cabo de Hornos* in Spanish are ridiculous. It is not a cape but a small uninhabited rocky island far off the South American mainland. It also has nothing to do with the English word *horn* or Spanish *hornos,* which means "ovens." It was discovered, in 1616, by William Schouten, who named it after his birthplace, the city of Hoorn, a seaport on the Zuider Zee in Holland. The English and Spanish names are due to misunderstandings and folk (i.e., false) etymology. Incidentally, Abel Tasman, eponym of Tasmania and of the Tasman Sea, also came from Hoorn, Holland.

§ Apropos of Bellingshausen versus Bellinggauzen, the Russian (Cyrillic) alphabet is more extensive than ours but it lacks letters equivalent to our *j* and *h*. In names and words foreign to them they use the equivalent of *dzh* for *j* and hard *g* for *h*. They spell my first name with the Cyrillic equivalent of Dzhordzh. One Russian author, evidently familiar with the pronunciation but not the English spelling of my name, gave it in Latin letters as J. G. Simpson.

§ It annoys me that some otherwise almost literate people speak of the "Artic" and the "Antarctic." Not only is that sloppily incorrect; but also it misses an interesting point. The north star is in the constellation of the bear, called *arktos* by the Greeks. Hence the far north was called *arktikos* in Greek, *arcticus* in Latin, *arctic* in English. The antarctic is anti-arctic, opposite to far north, hence far south. There are no such words as "artic" or "antartic."

§ For unfathomable reasons the inhabitants of Tierra del Fuego call their beeches *robles,* which means "oaks" in proper Spanish, including that of other parts of South America. The proper Spanish word for "beech" is *haya*.

§ Internal flights in Argentina are by the government air line. Unless things have bettered since 1972, that service leaves almost everything to be desired. Almost, not quite, as they do usually take you where you want to go. Perhaps they should also be credited with a first for me: en route from Mendoza to Tucumán my breakfast was a shot of Industria Argentina whiskey and a dry sandwich apparently several days old. (A double first, in fact, and last, for otherwise I have never had either a stale sandwich or distilled liquor for breakfast — champagne, yes, on rare occasions and only because it was there, as I am not particularly fond of champagne, but liquor, no, not before 11:30 A.M.) On the previous flight, Buenos Aires to Mendoza, the muddle at the in-town airport was even worse than usual, and the usual is awful there. I had no time or attention to examine our baggage checks carefully. When I did so, after departure, I was appalled to see that they had checked our luggage to San Rafael, not to Mendoza where we were going. I indignantly called a steward and insisted that they radio to San

Rafael and have things forwarded to Mendoza as soon as possible. He said in Argentine Castellano the approximate equivalent of "Cool it! All the luggage was put on this plane and as we do not go to San Rafael it will be put off at Mendoza and you can claim it there." For some reason, perhaps weakness after the struggle at the Buenos Aires airport, this struck us as so funny that we giggled all the way to Mendoza, where we found that the steward was right.

§ The name Falkland used by English speakers and by all inhabitants of the islands was given in the seventeenth century in honor of Viscount Falkland, one of a number of noblemen whose names have obsequiously been given to lands they never saw. Of course it is also quite sure that St. Malo never saw the Malvinas, at least not while he was on earth.

§ We thought the name Carcass Island rather macabre, but learned that it was not named for a carcass but for the sloop *Carcass* that visited the Falklands in the eighteenth century. I don't know why the ship was so named but note that in the eighteenth century the word *carcass* meant not only a body but also a sort of firebomb that could be discharged from a ship.

§ There were really worthwhile bergs as we worked back north from Adelaide Island, and I found myself mentally parodying a song of the gay 90s about an iceman:

Oh the iceberg is a nice berg
As I've learned once or twice,
But all I can get from the iceberg
Is ice, ice, ice!

— which is not quite true, because I also get a great deal of esthetic satisfaction from those sculpturesque phenomena.

References

Bridges, E. L. 1949. *Uttermost part of the earth.* New York: E. P. Dutton. I had read and greatly admired this book about early days in Tierra del Fuego long before I visited Ushuaia and became acquainted with Mrs. Goodall.

Holdgate, M. W., ed. 1970. *Antarctic ecology.* London and New York: Academic Press. Technical, but invaluable for anyone deeply interested in the Antarctic.

Pettingil, E. R. 1960. *Penguin summer.* London: Cassell. A delightful book about studying penguins in the Falkland Islands.

Quartermain, L. B. 1964. *South from New Zealand: an introduction to Antarctica.* Wellington (New Zealand): Government Printer. The best brief, popular introduction to the subject that I know.

Simpson, G. G. 1976. *Penguins: past and present, here and there.* New Haven and London: Yale University Press. I cited this before, but it is again relevant here.

Stonehouse, B., ed. 1975. *The biology of penguins.* London: Macmillan. A collection of somewhat technical chapters, but full of interest and not too difficult for many nonscientists; my summary of knowledge of fossil penguins is chapter 2 p. 19–41.

Strange, I. J. 1972. *The Falkland Islands.* Newton Abbot (England): David and Charles Harrisburg (U.S.A.): Stackpole Books. A fascinating book by our guide while we were in the islands.

Watson, G. E. 1975. *Birds of the Antarctic and sub-Antarctic.* Washington: American Geophysical Union. A standard work, more informative than the usual field guides for identification by birders.

23

A Dull Chapter

Dulness! Whose good old cause I yet defend. — Alexander Pope

A literary critic recently wrote "G. G. Simpson seems incapable of writing a dull book." His judgment is obviously superb, but one may question his reliability as he surely had not read all my books. If I put my mind to it, I can write as dull a book as the next fellow. I hope that this book can stand one dull chapter, a run-through of things that should enter into the record but for which time and space do not permit more than summary treatment. The events were not at all dull to me, some were indeed highly exciting, but quick enumeration can hardly be so.

Next are notes on some fairly recent travels not previously treated. They are here in chronological order. To some extent the chronology overlaps a few things that have already been discussed.

Until the circling of Africa in 1967, previously recorded, I had never taken an extensive trip that was purely for recuperation and vacation. In 1969, also with Anne, I did so again, this time to Alaska. We flew to Anchorage via Seattle and visited the Matanuska Valley and Katmai. What amused us most was the fishing procedure of a large brown bear near the Katmai camp: he lurked in the bushes until a human fisherman caught a salmon, and them preempted the salmon by right of ferocity. We went by train to McKinley Park, to view that great mountain and also, early one morning, a caribou stag with tremendous antlers silhouetted against the sky on a mountain crest. On by rail to Fairbanks where we visited the university and toured the Tanana River. At an island on that river a beautiful Eskimo girl demonstrated moose-calling. (You call them by shouting, "Moosey! Moosey! Moosey!" through a birchbark megaphone.) Plane to Whitehorse and train to Skagway, one of the finest rail trips anywhere. Ship to Vancouver with stops at Juneau and Ketchikan, the latter on Anne's birthday (20 August), and I bought her a necklace and bracelet of fossil walrus ivory. (Trade in Recent walrus ivory should be discouraged.)

In chapter 19 I mentioned the latter, African, part of our lengthy travels in 1970. Those travels began with a grant from the Royal Society to study fossil penguins in London and in Stockholm. Although it was not really unique for an American to receive an English research grant, this surprised American colleagues, who thought such grants were always in the reverse direction. At the British Museum (Natural History) I studied an important collection of fossil penguins from Seymour Island, just off the tip of the Antarctic Peninsula, and a small lot from Patagonia. I also took part in a symposium on early (Mesozoic) mammals at the Linnean Society and delivered the concluding remarks, or one might say the coup de grâce. In Stockholm I studied the rest of the then known fossil penguins from Seymour Island and also visited with colleagues, including old friend Erik Stensiö. That ended my grant project and we took off for the Indian Ocean and Africa, both enjoyable and instructive but self-indulgent rather than professional. I may say that since I left the American Museum all my travels, even those largely professional, have been at my own expense, excepting only the 1960 visit to Africa, when I had a Guggenheim Fellowship, and the 1970 visit to London and Stockholm, on a Royal Society Grant.

In 1972 I went with Anne to an annual meeting of the American Psychological Association in Honolulu, where Anne received an award "for her distinguished contribution to clinical psychology." We also treated ourselves to a tour of the other islands, having hitherto been on Oahu only.

In 1973 we were back in London for another symposium, this one at the Zoological Society on the hystricomorph (porcupinelike) rodents. I acted as chairman of the first session and made some remarks. We also visited friends, making our last visit with Julian Huxley, who had had a stroke but had made a partial recovery (he died in 1975). After the meeting we flew to Athens and then by ship toured the Greek islands, were much taken by Knossos and by the old walled part of Rhodes, visited Ephesus, by which we were properly but not unduly impressed, and Istanbul, where we admired Hagia Sophia, the Blue Mosque, and the Seraglio Museum. We were amused to see the filming of *The Abduction from the Seraglio* there. We disliked the rest of the city.

Later in 1973 we visited Yucatan and Guatemala, particularly to see the Mayan ruins, which had interested me since childhood but which I had not previously seen. My uncle Charlie Baldwin, husband of my father's sister Lil, had a copy of *Incidents of Travel in Central America, Chiapas, and Yucatan,* by John L. Stephens, and I read and reread it as a child when I spent two summers on the Baldwins' farm in North Carolina. Its two volumes were my (sole) inheritance from Aunt Lil and Uncle Charlie, and I still have them. I again read them after visitng some of the sites described by Stephens. It was hard luck that he missed Tikal, my favorite of all the Mayan ruins.

From Guatamala we flew to Guayaquil, Ecuador, and then to Baltra Island in the Galápagos. There we boarded a small but reasonably comfortable cruise ship, formerly the *Lina A* but rechristened *Floreana.* (She later sank, but without loss of life.) I had an intense Darwinian interest in the islands and also a specific desire to see the Galápagos penguins, which belie the penguin sterotype by living right on the equator. I became extremely ill, but thanks to the cruise personnel, who literally carried me off and on the ship, I did see a number of the penguins on the west side of the largest island, Isabela (the same as Albemarle; all the islands have both Spanish and English names). One of the guides even took me to an area where the penguins were nesting in holes in the lava and which is more or less off limits to tourists. When we finally, with difficulty, got back to Tucson it was found that I had had, and in part still had, lobar pneumonia and had survived only because an orthopedic surgeon, Dr. Drone, a passenger on board, had given me antibiotics that happened, really by chance, to be appropriate for my infection. I was convalescent for some time. Nevertheless, I was now well acquainted with all available fossil penguins and had a nodding acquaintance, at least, with twelve of the species of living penguins in their natural habitats. I started on a book to include both but especially the recent forms.

Our vacation trip in 1974 was a tour, or rather two end-to-end tours, on the *Explorer* to Arctic Canada and Greenland. We flew via Montreal to Frobisher Bay, which is in the Arctic tundra although south of the Arctic Circle, and then to Greenland's main civil airport. That is at an isolated spot near the head of the Søndre Strømfjord. There we boarded the ship. After taking on fuel and water farther down the coast, we crossed Davis Strait to the Canadian side, worked around Baffin Island into Lancaster Sound and Barrow Strait, where we had some ice problems, and up the east coast of Ellesmere Island to our farthest point north, in the Kane Basin at latitude 79°11' north. Back across Smith Sound to the former site of Etah, once the world's northernmost settlement, famed in polar exploration and still active in crossword puzzles but now deserted. Thence we returned down the west coast of Greenland to Søndre Strømfjord with several stops. Among them was the settlement, the northernmost, since Etah was abandoned, called Thule by Europeans and Qânâq by the Eskimos. These, the famed Polar Eskimos, were ousted from the original Thule-Qânâq about seventy miles farther south so that an American air base could be established there. On the map that is now "Dundas Vejrstation," as if every little Russian did not know that it is the American military base Thule. The Polar Eskimos now wear Danish sweaters, eat Tasmanian apples, and travel by sea in dories with outboards and by land in snowmobiles, although they keep huskies as pets and also as status symbols, much as the Navajos now all travel by car but keep as many horses as possible. The town of Jakobshavn, one of the largest in Greenland, boasted of having 200 Danes, 2,800 Greenlanders (Eskimos of former tribes farther south than the Polar Eskimos), and 6,000 dogs. The equivalent of a one-horse town in Greenland would be a one-dog town, but none such exist.

Our continuing cruise, or technically (and financially) speaking the second cruise, took us back to Canada and up Cumberland Sound to Pangnortung on the Cumberland Peninsula of Baffin Island. Here and elsewhere we were disappointed to find that it has become impossible to obtain good Eskimo carvings directly from the Eskimos. All are committed to commercial trade and most of them reach retail customers only in the big cities of southern Canada and of the United States. We went around the eastern end of Baffin Island, through Hudson Strait, as stormy as the Drake Channel, into the northern part of Hudson Bay as far as Southampton Island, then doubled back with various stops both on the islands and on the extreme northern part of the mainland (Québec) in this region, and finally reached Frobisher Bay where we disembarked and started the series of flights back to Tucson.

The Arctic and the Antarctic are similar in some respects, but on the whole are so different that polar experience requires seeing both. The tundra, absent in the Antarctic, has a unique and fascinating ecology and is especially attractive when in blossom. There are no flowering plants in the Antarctic. Both polar regions have millions of birds, and while a few are similar or even virtually the same (skuas), most are different: in particular, penguins in the Antarctic and the equally abundant, more varied members of the auk family in the Arctic. Seals occur in both, but they are more varied and more common in the Antarctic. Walruses occur only in the Arctic and they were for me the most fascinating denizens of that region. I was enraptured by hundreds seen on and around a big rock appropriately called Walrus Island in Fisher Strait between Southampton and Coats islands. Native land mammals, absent in the Antarctic, are abundant in the Arctic, notably the semiamphibious polar bears, a number of which we saw at close quarters, but also many lesser and more strictly terrestrial species. Icebergs are much the same in both polar regions, although, with the local exception of one burgeoning of bergs, those we saw in the Arctic were less abundant and less showy than those in the Antarctic. The geology of the Arctic is varied, interesting, and generally more accessible to view outside the ice caps. The scenery is grand in both, but within our experience more so in the Antarctic. The Antarctic has no native humans. Those of the Arctic are still interesting, but now almost incomparably less so than they appear in early accounts.

In June and early July of 1975 we went on the *Sagafjord* for a nominal North Cape Cruise. With some variations, that has been one of the world's standard cruises and with the usual exceptions, most of the several hundred passengers were cruise bums and not congenial to us. The food, service, and management were mediocre, but we did greatly enjoy most of the stops. Those included Amsterdam, Hamburg (where we spent all our time ashore at the great Haggenbeck zoo), and Copenhagen (where we made the trite visit to Tivoli but also spent some time watching a truly comic marionette show and truly handsome topless young ladies sunning themselves in another park) — cities we had not previously visited. From Oslo started the usual voyage up the coast and into several fjords,which we found just as striking as have some tens of thousands of other tourists and which I forbear describing. That part of our tour was unusual in only two ways: (1) a slight error in navigation caused our ship's superstructure to break a high-voltage power line, bringing it crashing and sparking down on our decks, and (2) because of bad weather and high seas our North Cape Cruise did not reach the North Cape.

However, after the capeless Cape tour we went on to Iceland, which was extraordinarily interesting to me in several ways. There is the Gull Foss, a waterfall not quite of world class but still excellent. There also is the prototype of all geysers. It has a proper name for itself alone, Geysir, so spelled and pronounced gay'-ser in Icelandic, elsewhere transformed to a common noun spelled geyser, pronounced ghee-zuh in England, where there are no geysers and the word designates shilling-in-the-slot water heaters, and guy'-ser in the United States and New Zealand, where there are geysers. The good old original Geysir now hardly ever spouts, but only a few meters away is a geyser that does spout handsomely at intervals averaging only four minutes or so.

Even more interesting for a geologist is the fact that Iceland is on the seam between the great crustal plates on which North America and Europe ride. Here rising from great depths come molten lavas that add to the plates on each side and slowly, inexorably move them apart, a process that started by separating them almost fifty million years ago. In most places where this process is active its loci are in oceanic depths, but here in Iceland clear evidence of it is well exposed on land.

I do sometimes have some rather profound thoughts but by now the reader knows that I also have some out-and-out trivial ones. In Iceland I was again struck by the well-known, indeed obvious, fact that a first essential for foreign travel is to distinguish signs for men's and women's toilets. In Iceland these are respectively *Karlur* and *Konur* words not in the vocabularies of most travelers. *Karl* is fairly easy from acquaintance with the sagas (-ur just indicates the plural), although we usually think of karls as low-class men and thus not really including ourselves. *Kon* is baffling and seems to have no cognate in English and only doubtfully some in other Scandinavian languages. The word for "woman" in Norwegian, for instance, is *Kvinne*, which is cognate with English *queen* but obviously not with *woman*, and not clearly so with Icelandic *kon*.

Late in 1975, or more exactly from 28 August to 30 September of that year, Anne, my sister Marty, and I flew to Manila and back, and in between took a voyage that I entitled "After You, A. R. Wallace, or as the salesman put it, 'Isles of the Orient Seas.'" In simpler terms, we toured the Philippines and Indonesia in a small, poorly maintained, badly manned, and ineptly run ship that had been retired from the California-Alaska run, bought by the Philippine government, and renamed the *Doña Monserrat*. To give blame where blame is due, the ship was not operated by the government but by a private Philippine navigation line based on Iloilo and was cruising not for that company but on charter for a travel agency based in Florida, U.S.A. To give credit where some little credit is due, we did have a pleasant and helpful cabin steward, Val Labra, from Iloilo, who spoke both the language of that island, Ilongo, and Tagalog, a very different Philippine language, fluently and had barely but sufficiently functional English. About seventy different languages are spoken in the Philippines, but the official language, which is called *Pilipino* (there is no sound *f* in the language) is a somewhat brushed-up version of Tagalog (accented on the *ga*), the language spoken in and around Manila. Indonesia is even more heterogeneous and has an even larger roster of languages but schools are uniformly taught in Malay, or at least in an only slight modified version of that language that is of course given the more patriotic name *Indonesia*, which means either the country or the language.

This travel through the Philippines and Indonesia was for enjoyment and relaxation, which was achieved in spite of drawbacks, but also had for me a more serious purpose. One of my interests almost throughout my life has been biogeography, the study of the past and present distribution of organisms on the earth. I have already indicated that in connection with South America. I had also made extensive studies of faunal exchanges and relationships between North America and Eurasia, and given some but less attention to African biogeography. My interest in the Australian fauna has also been mentioned; it, too, was in part biogeographic. In that connection and as a result of considerable reading, starting with Wallace, I had become especially concerned with the faunal relationships of the innumerable islands between Australia and Asia, most of which are now politically in Indonesia. Of course I did not expect in less than a month to make any important addition to knowledge of the already much-studied faunas of those regions. What I did expect and did accomplish was to see enough of environments, ecologies, topographies, and geographic relationships to make intelligent studies of the extensive, often confusing, and sometimes contradictory data and discussions already available. In the year following my return I made such studies and wrote a paper on the broader biogeographic relationships in this area.

In the midst of those serious considerations and of some fun with the Malay language, I had other amusements and as a sample I will share one with the reader.

One evening at dinner a couple at a nearby table managed to get a bottle of wine served by a new, handsome young Pilipino steward who had never served, perhaps never seen, a bottle of wine before. I transcribe what I heard the man at the table say and what I read in the steward's mind:

(*Pilipino to self*: Ladies first, ladies first.) *Begins to pour wine.*
Male passenger: Stop! Pour me a little first!

(*Pilipino*: Well, some tourists are not polite, or perhaps she doesn't drink. A Muslim, perhaps. So I'll fill his glass for him.)

M.p.: No! No! Just a little bit. A taste!

(*P.*: If he only wanted a taste, why did he get a whole bottle?)

M.p.: All right. You can pour it now.

(*P.*: So it is true that one sip can make a drunkard. All right, here goes.)

M.p.: No! Stop! *Her* glass now.

(*P.*: So, he finally remembers his manners.) *Fills her glass to the brim and starts to leave table.*

M.p.: Come back! Fill mine now.

(*P.*: All right. He may as well be drunk as the way he is.) *Fills glass, starts to leave again.*

M.p.: Hey! Leave the bottle here.

(*P.*, leaving: These Americans are strange enough even when sober.)

It happens that that was the same day a news bulletin from Manila included the statement, "U.S. officials said Kissinger may well not shepherd into reality in the manager of his 136 days hoping between Jerusalem and Alexandria to fix the Israeli/Egyptian accord."

That reminds me that my Tagalog dictionary-phrase book is full of extraordinarily bloody-minded illustrative phrases. For example, to illustrate the word *sapa*, "pool," which would seem innocuous at worst, the author has *May taong nagtatapon ng basura sa sapa*, "Some people throw garbage into the pool." That is mild for her. We made a game of constructing stories by stringing her phrases together. "Hypocrisy is one of her bad qualities." "He calls her an idiot." "She does not know that her marriage is illegal." "He has two illegitimate children." "This year he will be eighty years old." "He will die of hunger." "His wife drank poison and she died of it."

That must stop or some critic will think that this is a frivolous book, and another will remind me that I promised to make this a dull chapter, but not just this way.

Besides a number of islands in the Philippines, we visited Ternate (where Wallace had his base for some time), Banda, Ambon, Alor, Sawu, Flores, Komodo, Sumbawa, Bali, Java, and Celebes. What marvelous names! All were interesting; most of them were lovely. Bali could have been one of the lovely ones, as by all accounts it used to be, but now it is one, big, teeming tourist trap with roads lined by persistent brats holding up junk and yelling "One Dolla! One Dolla!" On

Java we saw, among other things, the great Buddhist monument of Borobudur, the height of one of my ambitions now that Angkor Wat cannot be visited by Americans. Java still is one of the lovely islands, although it has probably been inhabited by man as long as any place on earth — we passed near the place where the first *Pithecanthropus* (now called *Homo erectus*) was found. Mankind does not necessarily destroy its own environment. That is the good news. The bad news for me was that at Bandanaira in the Banda Islands my own awkwardness and that of a boatman reinjured my bad leg, so that I am now more dependent than ever on my large collection of canes, augmented by a nice carved one from Bali.

We have some rather complicated plans for both work and relaxation in the next year or so, but as of now (I am writing in December 1976) our most recent extended travel was another visit to the South Pacific and thereabouts in July and August of 1976. We flew out to Sydney where I had pleasant visits with Alex Ritchie, vertebrate paleontologist at the Australian Museum, and Charles Birch, geneticist at the University of New South Wales, as mentioned before. On 28 July we —Anne, Marty, and I — sailed on the *Fairstar,* an old ship not luxurious; but kept in excellent condition and well managed. It makes circular cruises, which may be begun or ended at any port, and its passengers are almost all Australians and New Zealanders, most of them on holidays. On our trip the great majority were Australians and the rest all New Zealanders except for one American couple who had lived in Australia for years and the three of us. Everything was fully informal and much of the action was in a bar fitted out like an Australian pub and serving oceans of beer. Most of our contacts in Australia had been academic and this crowd was of a sort less familiar to us — noisy, rowdy, having a great time in a cheerful and friendly way. The dining-room steward apologized at seating us with an elderly and at first sight somewhat difficult looking Australian woman, and he promised to give us a separate table for three at the first opportunity. We soon found our table-mate delightful and became very fond of her. When the time came to part she asked Anne's permission to kiss me good-bye, which I consider a fine compliment.

At Brisbane, our first stop, we had old friends and spent a long, extremely pleasant day with them. Toward the other end of the voyage we were quite familiar with Suva and had a good day even though we do not have personal friends there. In Auckland, where we left the ship, I spent some time at the university and gave a lecture there, and on another day we were taken on a long drive through a reserve much of which is almost virgin rain forest with kauris, native palms, groves of graceful tree ferns, and cabbage trees (usually called *ti* trees in other areas).

Between Brisbane and Suva we were in territory previously unknown to us: Lae, Wewak, and Madang on New Guinea (politically in Papua New Guinea), Rabaul on New Britain (also politically in Papua New Guinea; its harbor is named for some Simpson, unknown to me), Honiara on Guadalcanal in the Solomon Islands, and Lugenville (usually called *Santo*) on Espiritu Santo in the New Hebrides. Much of the touring was devoted to relics and scars of the war against Japan, which we found distressing, but we did manage to see some fine scenery and something of the now increasingly Europeanized natives, all quite dark Melanesians (*Melanesia* is from Greek roots for "black islands"). The Australians, who had been heavily engaged in the Pacific war we incline to think of as solely ours, had in some cases more reasons to visit war sites and cemeteries, as did one man who came along to try (successfully) to find the grave of his "mate" (buddy) — he said "mite" of course. The language of these middle-class Australians is sometimes difficult and does have some resemblance to cockney but is not the same. That reminds me that there were Aussie biddies on the ship and of a short conversation with one as we were about to go ashore on a very Melanesian island:

Biddy: Ah they dock hee-yah?
I: I beg your pardon?
Biddy: I sigh, ah they dock?
I: Oh, yes, they are dock, most of them.

We flew home from Auckland.

So much for travels. During these most recent years I was spending even more time than before on academic matters, research, writing, and occasional lecturing. I have already mentioned some of the research and publications of the late 1960s and the 1970s so far. Those and others included a study of community evolution for a Brookhaven Symposium, the argyrolagid monograph, a number of other studies on fossil marsupials and on fossil penguins, assurance (written jointly with Anne) to a conclave of psychiatrists or shrinks that man is not a naked ape, a chapter on the evolutionary concept of man for a book celebrating the centenary of Darwin's *Sexual Selection and the Descent of Man,* a summary of the origin of mammals published (in German) in *Grzimeks*

Tierleben, consideration of the concept of progress in organic evolution for an issue of *Social Research* devoted to ideas on progress, my second Burg Wartenstein Symposium paper, an article on Darwin published (in Spanish) in *Universitas,* the book on penguins, and much else, including a miniautobiography entitled "The Compleat Palaeontologist?"

References

Dickerson, R. E. 1928. *Distribution of life in the philippines.* Manila: Bureau of Printing. Out of date, but still basic for its subject.

Irving, L. 1972. *Arctic life of birds and mammals including man.* New York, Heidelberg, Berlin: Springer-Verlag. Perhaps attempting to cover too much for one book, but a good, somewhat technical compilation.

Kermack, D. M., and K. A. Kermack, eds. 1971. *Early mammals.* London and New York: Academic Press. Text of the symposium we attended in 1970; my contribution is on pages 181–98.

Rowlands, I. W., and B. J. Weir, eds. 1974. *The biology of Hystricomorph rodents.* London and New York: Academic Press. Text of the symposium we attended in 1973; my remarks are on pages 1–5.

Simpson, G. G. 1977. Too many lines. *Proceedings of the American Philosophical Society,* 121:107–20. This is the study of the zoogeography of the Indonesian islands written after my visit there.

Stefansson, V., ed. 1947. *Great adventures and exploration from the earliest times to the present as told by the explorers themselves.* New York: Dial Press. Covers much else, but includes 238 pages on Arctic exploration with excerpts selected and introduced by one of the great explorers himself.

Stephens, J. L. 1841. *Incidents of travel in Central America, Chiapas, and Yucatan.* New York: Harper and Brothers. A classic, still well worth reading, and with magnificent engravings by Catherwood.

Wallace, A. R. 1869. *The Malay Archipelago.* New York: Macmillan. For me, one of the three greatest books on travel, the others being Darwin's *Voyage of the Beagle* and Bates's *Naturalist on the River Amazons*; Wallace was our retrospective guide as we moved among the islands.

24

The Way Things Are

In real life serious things and mere trifles, laughable things and things that cause pain, are wont to be mixed in strangest medley. – John Keble

The author of that epigraph, almost the last for this book, at least, was known to me by name while I was still a young boy, for he rated three columns in the eleventh edition of the Encyclopaedia Britannica, the last erudite edition and my almost lifelong mentor on everything dating from the nineteenth century or earlier. Keble (1792–1866) was called "so simple, so humble, so pure, so unworldy" by his anonymous Britannica biographer, adjectives to which I have not even aspired, still less attained, and he is credited with founding the Oxford movement, with which I have little sympathy. Yet Keble expressed himself beautifully in English and, so we are given to understand, also in the colloquial Latin of his Oxford lectures. As I approach the end of my unconventional medley of serious things and mere trifles I find no more appropriate epigraph.

A long and active life eventually involves a cast of thousands of thoughts, experiences, and people. Obviously I have mentioned only a tiny fraction of those. Many of the people, although of special interest to me in one way or another, had no real part in or influence on my own life. For example, as a small boy I attended the burial of Buffalo Bill (Cody) and was puzzled by the fact that there seemed to be two widows at the graveside; and at a different level of notoriety or fame, as an adult I have talked briefly, at least, with several presidents of the United States. Most of those whom I have mentioned in previous pages have had definite roles in my life, and I in theirs, in many instances as professional colleagues in various fields. Some who have had large but nonprofessional parts in my life, even very affectionate ones, have not been mentioned. Here I should at least name Creighton B. Peet, until his death in 1977 a close, valued friend for more than half a century, his wife Ann, and their son Creighton H. Peet, to us almost the son that we have not ourselves had.

The miniautobiography that I mentioned at the end of the last chapter and will cite at the end of this one is quite different in style and content from the present maxiautobiography. It was written as a chapter for a technical annual publication. It relates almost no specific incidents in my life but, as suggested by the editor, summarizes some of the changes that have occurred in a few fields of my activities during some fifty years of professional life and suggests my present judgment of the status of those fields.

The present informal, less serious, less technical, more narrative, and much longer account has now made sufficiently evident in general what I have done with my life, why, and with what result. I have collected a great many fossils, described even more, and named a good number of them. That, together with the relevant geological observations in the field, constitutes the basic, observational, and relatively objective part of my special science or subscience, vertebrate paleontology. Beyond that, I have taken a broader stance and a more theoretical and subjective one always in part in geology but increasingly also in organismal and evolutionary biology. Associations with colleagues have been almost evenly divided between geologists and biologists. In the present, late phase my local professional association is almost entirely in geology, although some of my recent work is more biological.

In the miniautobiography the subjects treated most specially, in eight out of eleven text pages, are evolution, systematics, and biogeography. All three have repeatedly appeared in appropriate parts of the preceding narrative. I will now state, more briefly than possible, one might say, my present judgment of these three enormous subjects.

As what is now called the synthetic theory of evolution began to develop, soon after I was out of graduate school, I early espoused it. I made some substantial contributions to it, starting long ago, so that I believe it is not unduly immodest to number myself among its several cofounders. It is now accepted in its essentials by a clear majority of organismal biologists, that is, those whose studies involve whole organisms and aggregations of organisms rather than, or in addition to, the chemistry and physics of parts of organisms. With accretion from many sides this body of theory has become increasingly broader and also more intricate. Many details are still uncertain and no one supposes that a complete explanation of evolution has yet been attained. A really *complete* explanation may not be possible for this, or perhaps for anything.

Some dissent in recent years seems to me to be apparent rather than real. In some instances what is involved is only a restatement in different terms, or with some change in emphasis, of points made in essence, at least, long before. I believe, for example, that this is true of the theory of *punctuated equilibria* now under considerable discussion. In other instances what is new is a reformalization or extension of elements inherent in the body of theory. I believe that is true, for example, of the theory of island biogeography, previously mentioned and now involved not only in that special aspect of biogeography but also in broader matters of evolutionary theory. Still other recent contributions are indeed quite outside the previous body of theory but do not contradict any aspect of the synthetic theory and thus can, and I am confident will, become an extension of that theory, which is not bound to a static dogma. For example, I believe that to be true of the possible role in evolution of recombination of genes from widely different species.

The most formidable fairly recent development that is possibly but not clearly consonant with the synthetic theory is the hypothesis of non-Darwinian evolution, incorrectly so called. Advanced by biochemists, not organismal biologists, this holds, in simplest and most general terms, that many, most, or possibly even all genetic changes in populations (or taxa) are not affected by natural selection and occur at random and at constant rates. The synthetic theory recognizes that mutations are essentially stochastic, that is, are statistically random within certain definite constraints. It holds, however, that any phenetic effect, that is, a definite change in the organism outside of the genetic system, is potentially subject to natural selection and is highly likely to become so eventually if not immediately. The only mutations that are likely to be selectively neutral are those with no direct or indirect phenetic effects, and if any such really exist (a point difficult to prove), they are of no interest for organismal biologists as they are not involved in the evolution of whole organisms. That fixation of mutations in populations occurs at a constant rate is demonstrably false and the theory that base substitutions in proteins, for example, provide evolutionary clocks is not tenable.

In chapter 17 I mentioned my book on classification and proposed to return to controversy that later developed in that field, which I now do. Although my 1961 book, written in connection with the Jesup Lectures, was my most extensive contribution devoted exclusively to principles in this field, it must be apparent that a great part of my research has always involved the practical application of such principles. It does also occur to me that some — I think not a majority — of those vociferous in discussion of the principles of classification are quite deficient in its practice.

Most of those primarily involved in the practice of classification and proficient in it have followed and exemplified the principles of evolutionary or, as I now prefer to call it, eclectic taxonomy. My main contribution was to make those principles explicit and well organized as of 1960. Eclectic taxonomy continues to be applied with increasing sophistication not so much in principle as in methodology. Since then two other schools, initially quite distinct in principle, have arisen. The first was a conscious revival of an eighteenth century, hence long preevolutionary, principle amplified and complicated in the computer age and in this avatar dating largely from a work by Sokal and Sneath published in 1963. They called it *numerical taxonomy,* but that is a bit snide since all modern taxonomy is extensively numerical. It is more distinctively known as *phenetic taxonomy.* In its original, purist form it rejected any evolutionary, phylogenetic, or genetic considerations and insisted, sometimes in objurgatory terms, that classification should be based directly and exclusively on degrees of physical, organic (phenetic) resemblance calculated and expressed numerically.

That is a perfectly logical and workable system, although initial claims that it is wholly objective are demonstrably false. In their own terms, classifications developed in this way are valid. The most basic objection to them is that they leave out of consideration genetic affinities and evolutionary principles that most taxonomists consider relevant to the most useful and desirable sorts of classifications. The pheneticists have developed elegant methods for computing degrees of resemblance, and there is an increasing tendency to welcome these not as a final end in themselves but as useful means toward classification when interpreted by genetic and evolutionary principles, thus depolarizing the situation and bringing numerical phenetics into eclectic taxonomy, another form of synthesis.

That is not true and perhaps will not eventuate with regard to another, currently controversial system of taxonomy. This was proposed primarily by a German entomologist, Willi Hennig, in 1950, but it received almost no attention until a revised, rewritten version was published in English in 1966. (There is internal evidence that this later

version was completed at least five years before publication and not further revised.) The main principles of the Hennigian system are: first,
that the basic process of organic evolution (phylogeny) is the splitting
(dichotomy) of an ancestral species into two descendant species; second,
that each dichotomy should be taken as marking the origin of two new
units (taxa) of classification; and third, that the hierarchic level of such
units (whether species, genera, families, etc.) should be determined by
the geologic time when the dichotomy occurred, the earlier the time,
the higher the level.

I have already mentioned that when a facetious remark I made in a
closed conference was played from tape before an unintended audience
a Hennigian in that audience had what our ancestors called a conniption
fit. I do not want readers of this book to have fits, so I will not now be
facetious. I will just say that the first principle, as given above, an
apparent statement of fact, is not true and that the second and third
principles, statements of opinion, are inane. Rather than arguing the
matter here, I refer anyone professionally interested to the discussion by
Ernst Mayr cited at the end of this chapter.

What I have said so far about theories of evolution and principles of
taxonomy is that, apart from matters of detail and taking into account
accretion and complication in recent years, the views that I supported
and expounded years ago are still basically valid and preferable to others
on the same subject. I am quite aware that this might be interpreted as
loss of mental flexibility in someone nearing the end of a long career.
My own judgment of that possibility is of course suspect, but there is
evidence to support me. First, a great many of my younger colleagues,
indeed a clear majority on the evidence available, agree with me in
those two fields. Second, since reaching my seventies I have myself
changed my mind on a basic principle of another field of major importance to me: paleogeography.

My early acquaintance with paleogeography was largely through the
writings of Charles Schuchert and eventually personal contact with
him. He had retired before I entered graduate school but was still very
much present in the Yale Peabody Museum, where I spent my graduate
years, and he had an open door and a cordial attitude toward students.
In spite of his keen intelligence and great authority in this field, he was
holding to some views even then becoming untenable, notably the
former existence of continents that foundered to form the present ocean
basins. I soon (in 1924) became acquainted with W. D. Matthew and
his different views. Some nine years earlier, in 1915, he had published

what became one of the most seminal or heuristic studies of paleogeography and historical biogeography, under the rather unrevealing title
of *Climate and Evolution*. Matthew was not so much concerned about
foundering continents, which he and many others held had never
existed, but about alternative hypotheses of multiple land bridges con-
necting areas now separate. Paleogeographers of that school had virtually enwrapped the entire earth in a web of such hypothetical bridges.
Although he covered many other points, the heart of Matthew's theory
was that the distribution of land animals throughout the relevant part of
geological history, and most particularly the distribution of mammals
in the Cenozoic or Age of Mammals, is better explicable if present continents and oceans had been essentially stable and connections between
continents had been those still existing and a very limited number
across what now are narrow or shallow waters and not oceanic basins.
A Bering Sea connection between what are now Siberia and Alaska was
the clearest example of such past connections.

It happens that almost simultaneously with Matthew's major publication on this subject, a quite different theory was expounded by a young
German meteorologist, Alfred Wegener, in a small book entitled *Der
Entstehung der Kontinente und Ozeane* ("The Origin of Continents and
Oceans"). The idea had been knocking about for a long time, but this
was its first complete and coherent statement. The theory is that the
present continental crustal segments once formed part of one or a few
supercontinents — Wegener himself had only one but a number of
adherents to the theory started with two — and that the supersegment
(we now call the segments *plates*) broke apart and the fragments moved,
drifted, slowly into their present positions as our continents.

The theory as Wegener advanced it rested on minimal evidence,
much or all of which could also be explained by other theories, including that of stable continents. Continental drift thus was highly controversial, and for a long time it was rejected by a clear majority of
geologists, including virtually all paleontologists. Wegener's theory was
not based primarily on paleontological and biological grounds, but he
argued at length that evidence from those fields supported the theory.
The theory thus became highly relevant to my increasing concern with
biogeography. I soon found (and this is still correct) that most of
Wegener's supposed paleontological and biological evidence was equivocal and that some of it was simply wrong. (Wegener had no firsthand
acquaintance with these fields.) Other early supporters of continental
drift, all nonpaleontologists, even more grossly misinterpreted the

paleontological evidence. The limit of absurdity was reached by a French geologist who claimed that North America and Europe had repeatedly drifted together and apart during the Cenozoic by a *"mouvement en accordéon"* (italics his) to account for supposed mammalian trans-Atlantic migrations that quite surely never occurred.

Thus in the 1930s and 1940s after lengthy investigation I concluded that the then–available real evidence of known land mammals not only did not support but opposed any effects of continental drift during the Cenozoic. I found that the known distributions of mammals during that era could be explained by even fewer land-bridge connections than Matthew had accepted. I did not deny the possibility of earlier effects of drift, but at that time I considered evidence for the drift theory so scanty and equivocal as to make it an unconfirmed hypothesis. I thus became known, not quite accurately, as a strong opponent of Wegener's theory.

In the 1960s and increasingly in the 1970s the situation has entirely changed. There has been a revolution, or at least a major subrevolution, in physical geology. By discoveries for which even the techniques did not exist during the long controversy over continental drift, it has been demonstrated that the earth's crust consists of a number of discrete areal segments or plates that have at some, but not all, times in the earth's history been in motion with respect to each other. The continents as we know them have ridden on these plates, and they have drifted when these plates moved. This is not the place to review the new evidence and I just mention that much of it involves the discovery that the earth's magnetism has changed through geological history and that its direction at various times in the past can be determined. Another body of evidence comes from exploration of the beds of the seas and discovery that seas spread from rift zones that are boundaries of plates and sources of additions to them.

Although I am still sometimes cited as an opponent of continental drift, I fully accept this great addition to our knowledge of earth history. It still is highly probable that continental drift had little effect on the geography of land faunas during most of the Cenozoic. The only important exceptions, in my present opinion, are that North America and Europe were evidently parts of the same continent until about the end of the early Eocene, when their drift broke the last connection, and that Australia was probably in the vicinity of Antarctica until some time in the early Cenozoic and did not reach approximately its present relationship to the East Indies and, through them, to Asia until some time in the Miocene.

So much, or so little, then for changes in some of the major fields of study in which I have been involved, and for my present position regarding them. Let me now close with a few more personal or at least less professional remarks on life in general and mine in particular.

What about those great questions that I tossed somewhat jauntily into the foreword of this book? I repeat them:

> What is the history of life?
> What are the causes and patterns of evolution?
> How has the history of continents affected the distribution of animals?
> What is the nature of the human species?
> What of ethics and gods?

The question now is, "Have I answered those questions?" Certainly not completely. Still more certainly not finally. Probably none will ever be answered in a truly complete and final way. Although I am an anti-fan of boxing, it occurs to me to comment on this in boxing terms. It is as if for each of those questions, and a great many more, I had entered the ring, fought my best, delivered and received many a hard blow, and achieved a tie. Good fights! I am happy that I held my own. I am also happy that the fight helped a bit to put each question in its place. Neither I nor the questions have retired as a consequence.

There is currently much discussion about the motivation of scientists and other professionals. The motivations are numerous and usually complex even for any one individual. Those of us so oriented ponder our own motivations. In the main, mine is surely that I was born with or somehow very early acquired an uncontrollable drive to know and to understand the world in which I live. I have not aspired to promotion and office; on the contrary I have refused opportunities for them. I have received many tokens of recognition such as honorary degrees, medals, and other awards, honorary memberships in learned societies, and other such things. I could not quarrel with anyone who felt that I have received more recognition than I have deserved, and I cannot deny that I enjoyed it. But I have never done anything motivated by desire for recognition and I have never sought it. I have suffered about as much over mistakes in my pursuit of knowledge and understanding as I have rejoiced in the apparent successes.

Outside of my profession I have had a great deal of physical pain and emotional sorrow. I do not regret that, but rather value it as a part of the complete fullness of life. It is part of my good fortune that the pain and sorrow are more than overbalanced by pleasure and delight. Here in the closing years of my life I have a loving and fascinating wife and family and many good friends. I am enjoying, more than ever before, the company of students and colleagues at the university with which I am still connected.

I have answers to the banal questions: "What would you want to do if you had your life to live over?" "Who would you be if you could be anyone on earth?" To the first question: "The same." To the second: "Myself." Is that conceit? I think not. I think it is evidence of a success-ful life.

كل‎   مبد ي‎   متقوم‎

"Everything begun is ended." – North African proverb.

Notes

§ By coincidence, when I was writing this chapter's brief comment on my part in the development of the synthetic theory of evolution I received a letter from a colleague who is working on a history of that theory. He suggested that I had not claimed sufficient credit for my contributions to it. Well, my contributions are on record, and I think their evaluation can be made more becomingly and more fairly by others than by myself.

§ I am not here citing the leading or founding works on phenetic and on Hennigian taxonomy, as I think their interest is now purely technical and will eventually be purely historical. Students who do have a technical interest can locate them readily enough. I have already cited a work on biometry by Sokal and Rohlf that expounds elegant numeri-cal methods without the propaganda for so-called numerical taxonomy.

§ Hennigian taxonomy is called phylogenetic taxonomy by Hennig and most Hennigians, but that is misleading. Evolutionary or eclec-tic taxonomy takes account of all aspects of phylogeny, and Hennigian taxonomy excludes important aspects of phylogeny. Hennigian taxon-omy is usually called *cladistic* by non-Hennigians, because it focuses attention on lines of descent, *clades*, and their supposed bifurcation, but eclectic taxonomy also takes account, and more reasonable account, of clades.

§ I was not present when the unfortunate Hennigian was so upset by the tape of my facetious remark about Hennig, but the description of what occurred suggests a good, old-fashioned conniption fit. The *Dictionary of Americanisms on Historical Principles* traces that lovely expression to 1833 when one S. Smith wrote that "Ant Keziah fell down in a conniption fit." It may have been a snit that my colleague suffered, but to my surprise a quick run through a few dictionaries does not find that word, even though I hear it in common use.

§ In the paper cited below Mayr emphasizes that the reconstruction of phylogenies, which have been facts in nature even though we are not often able to restore them with certainty, is quite different from the erection of classifications, which are artifacts even when made consistent with phylogeny. Still Mayr has gone a bit overboard in giving Hennig "great credit" for "cladistic analysis," that is, the reconstruction of phylogeny. Hennig has indeed emphasized and defined some procedures of phylogenetic analysis that have long been used by systematists. He might even have clarified them if he had not unnecessarily replaced the usual plain-language terms by a bizarre and idiosyncratic new terminology.

§ Matthew died in 1930. He never reverted to paleogeography and never published on Wegener's theory, although he privately referred to it as "very ingenious and plausible" but "quite impossibly wrong" in Wegener's treatment of land faunas — certainly a fair judgment.

§ My excursion into paleogeography was honored in 1944 by a fierce attack from the South African Du Toit, one of the most emphatic of early supporters of continental drift. As I was overseas on military duty, my colleague and former teacher at Yale, the late Chester Longwell, undertook my defense. The incident is treated at some length in Marvin's book, cited below.

§ Although I must give Wegener a failing mark in paleontology and biology he merits a very high mark as a brilliant and innovative geological theorist. It is sad that he died in 1930, at only fifty years. If he had not frozen to death on the Greenland ice cap he could have lived to see the triumph of his concept of continental drift.

References

Marvin, V. B. 1973. *Continental drift: the evolution of a concept.* Washington: Smithsonian Institute Press. An interesting, essentially complete, and fair history, including an account of my involvement in the controversy.

Matthew, W. D. 1915. Climate and evolution. *Annals of the New York Academy of Sciences* 24: 171–318. This work retains great historical importance; it is more readily available in a reprint with added matter issued in 1939 as volume I of the Special Publications of the Academy.

Mayr, E. 1974. Cladistic analysis or cladistic classification? *Zeitschrift für zoologischen Systematik und Evolutionsforschung* 12: 94–128.

Simpson, G. G. 1976. The compleat palaeontologist? *Annual Review of Earth and Planetary Sciences* 4: 1–13.

Postscript

Ἀλλ ἦ τοι μὲν ταῦτα θεῶν γούνασι κεῖται ["Indeed these things lie on the knees of the Gods"] — Homer, if he (or if you follow Butler, she) composed the *Iliad*

During one of my many summers spent hunting fossils in the San Juan Basin a learned Jesuit (not Teilhard) shared my camp. Whenever he spoke of the future, he would hurriedly append that line in classic Greek, pagan as it is, much as a devout Spaniard will hedge on any prediction by adding "Ojalá," which is a corruption of an Arabic invocation meaning "if Allah wills!"

The Homeric warning that mere mortals should not count on what the gods have in wait for them has stuck in my memory, and I quoted it when in first draft I ended the manuscript of this book with a summary of some plans that would modify a few things that I had written. After the preceding pages were edited but before they were in type, two of those plans were successfully carried out. Rather than revising the previous text, I am making the emendations, or additions, in this postscript.

I have mentioned that I had not seen the Himalayas and that my evaluations of the scenery of all the continents except Asia were conditional because they did not include the highest mountains of all. Now I have beheld them. In February and March of 1977 Anne and I flew to Calcutta with short stopovers in Washington and London, then from Calcutta to Bagdogra, the nearest airport to Darjeeling. We were driven up the long, long hill to Darjeeling and spent some time mountain-watching in and around that fascinating small city. Its backdrop is Kangchenjunga, which is the third highest mountain in the world and at 28,168 feet is only 860 feet lower than Mount Everest, for which I prefer the native name Chomolongma. Kangchenjunga towers some twenty-one thousand feet above an observer in Darjeeling, which is itself atop a mountain. Chomolongma is about 110 miles from Darjeeling and not visible from there, but we saw it at sunrise from a higher crest near Darjeeling and again from the air as we flew back from Bagdogra to Calcutta. From Calcutta we flew to Hong Kong and then to Tucson over the Pacific. Thus we circled the globe again, this time from west to east, and it is no longer true that we have only once literally gone around the world and only from east to west.

To me Kangchenjunga is the most magnificent mountain on earth, or mountain-massif with its surrounding somewhat lower peaks. More attention has been given to Chomolongma only because it is insignificantly higher and because, although Kangchenjunga has also been ascended, in recent years it has been easier to get permits to climb Chomolongma (from Nepal). More climbs on the latter have been made, and they have been written about at greater length.

In spite of my enthusiasm for Kangchenjunga, Darjeeling, and the great span of the Himalayan "Snows," I still think that the Antarctic Peninsula is even more scenic.

As an etymophilist I must add a note on the pronunciation of Himalaya. The British usually call it Him-ah-*lay'*-ah, despite the fact that the most august of all British dictionaries, the O.E.D., labels that pronunciation incorrect. So does the most authoritative of all American dictionaries, the unabridged Merriam-Webster, but some Americans, tin-eared like so many British, do use the incorrect pronunciation. The word is Him-*ah'*-lah-yah. It is derived from Sanskrit *hima-alaya,* "abode of snow." The first *a* in *alaya* is long and must receive the stress accent in the compound Himalaya, all the more so because it has also absorbed the *-a* of "hima." English-speakers in India also call the whole range of the Himalayas "the Snows," which is apt and is a translation of *alaya.*

Most of the people who heard that we flew all the way around the world just to look at a mountain or two decided that we were either crazy or lying. Well, we weren't lying.

In chapter 9 I briefly told of my futile visit to Moscow in 1934 and mentioned the negative outcome of recent invitations to revisit that city. Since I wrote that chapter I have finally been able to make such a visit, in May 1977, just forty-three years after the early debacle. This time I went as a guest of the Institute of Morphological and Evolutionary Zoology of the Academy of Sciences of the USSR. I was received with the greatest cordiality by colleagues in my host institute and also by many in other institutes of the academy, in the University of Moscow, in the Lenin Library, and elsewhere. This new experience was so pleasant that I considered modifying the sardonic account of my earlier stay in Moscow. But that account is truthful and could justifiably have been made even more sardonic, or downright angry, so I have left it as first written in this book. The great difference between the two visits resulted from a combination of causes, but especially from the fact that I met no colleagues in 1934 (the Academy of Sciences was then in Leningrad; it moved to Moscow in 1936) and I was dealing entirely with obstructive bureaucrats and as an unwanted wayfarer. In 1977 for the most part I was dealing, as a welcome guest, with nonbureaucratic, largely apolitical scientists.

Although the fieldwork of the Polish-Mongolian expeditions in Mongolia, mentioned in chapter 9, has come to a halt, that of Soviet-Mongolian expeditions there is proceeding on a grand scale. Discoveries include Mesozoic mammals, undescribed as yet but shown to me without restrictions. I also met two of the Mongolian paleontologists from Ulaan Baator, Drs. Dashzeveg and Barbold, able and attractive men whose parts in the research are real and not just diplomatically *pro forma*.

On this visit I also had an unplanned glimpse of Leningrad, and in Moscow (along with some tens of thousands of both Soviet and non-Soviet tourists) I was now allowed to wander about in the Kremlin. And I was treated to excellent caviar and superior vodka, but alas! it is still true that my visit in 1934 was the only time when I have had all the caviar I could eat. I will no longer predict, and I drank many a toast to my return to the USSR, but now, even more than before, the possibility of another visit there lies on the knees of the gods.

Regarding my many further plans, I will say only "If Allah wills" —

ان شاء الله

Index

Academy of Natural Sciences of Philadelphia, 142
Academy of Science (Norway), 153
Academy of Sciences (U.S.S.R.), 78, 279–80
Acclimatization societies, 232
Acts of the Apostles, 14
Adaptation, 113
Addison, Joseph, 19
Adelaide, Australia, 105
Adrian, Lord, 146
Agassiz, Alexander, 183
Agassiz, Cécile, 183
Agassiz, Louis, 183, 194
Age of Mammals (Cenozoic), 50, 52, 62, 66, 67, 79-80, 117, 166, 272, 273
Aguerrevere, Santiago, 85
Aguirre, Fr. Emiliano de, 187, 188, 189, 190, 191, 193
Alaska, 256–57
Albert, Prince Consort of England, 48
Alexandra, Princess, 224
Alhambra, 190, 195
Alice, Princess, 235
Alice Springs, Australia, 229, 235
Alioto, Joseph, 137
Allingham, Margery, 156
Amahuaca Indians, 169
Ameghino, Carlos, 50, 53, 57, 60–61, 62, 63, 64, 65, 69, 70
Ameghino, Florentino, 50–51, 56–57, 60–61, 62, 63, 64, 65, 69, 70, 72
Ameghino, Léontine, 50
Ameghino Collection, 50, 70
Americana, Brazil, 163
"American Mesozoic Mammalia" (Simpson), 17
American Museum of Natural History, 49, 131; Anne Roe and, 87; Fayum expedition of, 139; Simpson's career at, 33–34, 35, 38–42, 65, 95, 114, 116, 117, 128–129, 133, 170, 172, 173, 181, 184, 214, 219
American Philosophical Society, 118, 129

American Psychological Association, 177, 180, 181, 221, 257
American Psychologist, The (journal), 180
Amiel, Henri, 108
Ancient Mariner (Coleridge), 15
Andes, 94
Angel, Jimmie, 4–7
Angel, Marie, 4
Angel's Falls, 4, 102
Animal Species and Evolution (Mayr), 116
Anner, Anne Marie, 10–11
Anning, Mary, 41, 44–45
Anonymous (burro), 87, 90
Antarctic, 102, 237, 240–48, 259–60, 273, 279
Antonio (Kamarakoto Indian), 4–5
Arabic language, 154
Arambourg, Camille, 152
Arambourg, Mme. Camille, 152
Araxá, Brazil, 161–62
Archaeology, 118
Archer, Mike, 226, 227
Archivo de Indias, 189
Arctic, 252, 259–60
Argentina, 164–66, 190, 237–40, 253–54
Argyrolagids, 219, 265
Army, 38, 39, 121–26
Arnold, Matthew, 25
Astraponotense, 70
Atacama desert, 104
Atrichornis clamosus, 228
Attending Marvels (Simpson), 49
Attlee, Clement, 151
Australia, 105, 142–47, 222–31, 235–36, 264–65, 273
Australia and New Zealand Association for the Advancement of Science (ANZAAS), 142, 146, 154
Australian Museum, 142
Autobiography (Darwin), 112
Avila, Spain, 193
Ayala, Francisco, 193
Ayers Rock, 230